TWO PHASE FLOW, PHASE CHANGE AND NUMERICAL MODELING

TWO PHASE FLOW, PHASE CHANGE AND NUMERICAL MODELING

Editor

Jeff Winters

scitus academics

Two Phase Flow, Phase Change and Numerical Modeling
Edited by **Jeff Winters**

ISBN: 978-1-68117-658-1
Library of Congress Control Number: 2016946525

SCITUS Academics LLC,
616, Corporate Way, Suite 2, 4766,
Valley Cottage, NY 10989

www.scitusacademics.com

Notice

Printed in United States of America on Acid Free Paper

PREFACE

The phase of a matter can be a liquid, solid or gas. The multi-phase flow is the multi-phase simultaneous flow. In energy related industries and applications, the study of multiphase flows is very important. A two-phase flow is the simplest case of multiphase flow. Two-phase flows can be solid-liquid, liquid-liquid, gas-solid and gas-liquid. Heat transfer and analysis on laser beams, evaporator coils, shell-and-tube condensers, two phase flows, complex fluids, nanofluids and phase change are important issues in a number of industrial processes and equipment styles. The most common class of multi-phase flows are probably two-phase flows. A liquid-gas flow is probably the most important form of multi-phase flow and is commonly found in a wide range of industrial applications. These include oil gas mixtures, evaporators, boilers, condensers, submerged combustion systems, sewage treatment plants, air conditioning and cooling plants and cryogenic plants. In the meteorology and in some other natural phenomena, gas-liquid systems are also essential. The flow of solids suspended in gasses is essential for pneumatic transport and for pulverized combustion of fuel. Fluidized beds can also be considered a form of solid flow of gas. Liquid-liquid flows can be referred to the flow of oil-water mixtures in pipes and mass transfer systems for the extraction of liquid-liquid solvents. The hydraulic transport of solid materials is the most important application of solid fluid flow. Liquid-solid suspensions are also present in crystallization systems and hydrocyclones. Historically, the usual two-phase flow cases are likely to occur in large power systems. Gas fired power stations and coal used large boilers for turbine steam generation.

This book covers state of the art numerical heat transfer modeling; two phase flow, nanofluids and phase change. The flow of two liquids which is completely insoluble in each other is an instance of liquid-liquid flow, such as water and oil is significant in oil recovery procedures. The injection of water diminishes the resistance to flow and the pressure grade into the oil flowing in the pipelines.

TABLE OF CONTENTS

1

3D MODELING AND SIMULATION OF A TWO-PHASE MIXING JET FLOW

Jaci C. S. C. Bastos1, Udo Fritsching2 and Milton Mori1*

1 University of Campinas, Campinas, São Paulo, Brazil

2 University of Bremen, Institut für Werkstofftechnik, Bremen, Germany

KEYWORDS

Gas-solid flows; Two-phase jet; Computational Fluid Dynamics (CFD)

1. INTRODUCTION

Two-phase jet flows are extensively used in a variety of engineering application sectors; more precisely, fluid flows containing solid particles, such as chemical, pharmaceutical, healthcare, biomedical, fuel, personal products, minerals industries and new materials. In all these applications, a fundamental understanding of how particles interact with fluid flows is necessary to allow the use of computational fluid dynamics (CFD) models in the optimization and performance improvement of existing equipment and processes; the identification and solution of operating problems; the evaluation of retrofit options and the design of new equipment, systems and plants including process scale-up [4].

The dynamic behavior of a gas-solid jet is defined by the complex interaction between its individual phases. Previous research indicates that in this type of flow the effects of entrainment and mixture of large quantities from the jet outer boundaries in a direction to the jet core are observed. The radial flow is defined as having a dense central solids region with high velocities for both phases and a low solids concentration near the boundaries. Furthermore, coherent structures are responsible for transport of significant mass, heat and momentum without being highly energetic, typical structures in shear flow are originated from some flow instabilities, mainly in the case of free shear layers (Decker et al. [5]).

Hardalupas et al. [6] investigated the velocity and particle flux characteristics of a turbulent particle-laden jet and found that the mass mixture ratio has a little effect on the particle concentration distribution. Hadinoto et al. [7] investigated a downward flow of glass bead particles in a vertical pipe for different Reynolds numbers and a constant particle loading with two particle sizes. For the 70μm particles, the authors observed that the presence of the particles damped the gas-phase turbulence intensity for smaller Reynolds numbers. Then, these results were compared with the single-phase flow for the same Reynolds numbers. As a consequence, it was observed an enhancement of the turbulence intensity at higher Reynolds numbers for the twophase flow.

Multiphase flow equations have been developed and analyzed by many researchers, such as He and Rudolph [8], Theologos and Markatos [9] and Ali and Rohani [10]. However, Soo [11] is credited with the mathematical approach to this type of flow. Rietema and van der Akker [12] presented a detailed derivation of the momentum equations for disperse two-phase systems.

Currently, two of the most well-known methods for the mathematical modeling of the solid-phase in numerical simulations of gas-solid flows are the discrete particle simulation and the two-fluid approach. In both approaches gas-phase is described as a local average of the Navier-Stokes equation and both phases are usually connected by a drag force. At first, the dense gas-solid problem could be resolved using just the Newton motion equations for each suspended particle and the Navier-Stokes equation for the gas phase (Fairweather and Hurn [4]).

However, due to the large number of particles, the resultant number of equations would be too large to allow a direct solution, at least with the computer capacity currently available. Thus, the solid-phase is also treated as a continuous-phase, exposed to the analogous conservation equations for the fluid phase. The Eulerian-Eulerian approach proved to be capable of predicting the gas-solid flow, as seen in the research conducted by Meier and Mori [13], Alves et al. [14] and Bastos et al. [15] and it was used in this study.

The specific interest in the present research is flow simulations containing a particle-laden jet. Researchers have shown that the gasphase and the solid-phase axial and radial distribution can be computed using the two-equation and zero-equation respectively, based on multiphase models. In general, the performance of the current models critically depends on the accuracy of the drag force formulation, which has been extensively analyzed in the studies of Gidaspow [2], Meier and Mori [13], Alves and Mori [14], Meier and Mori [16], Alves et al. [17], Bastos et al. [15]. However, as the drag force is described by empirical models, it is of extreme importance to evaluate them in accordance with the equipment employed. Due to this criterion, it was chosen the gas-solid flow jets to study the influence of loading particles in the mixture on its fluid dynamics.

The turbulent regime in studying is mainly dependent on momentum conservation. This enables us to consider separately the influence of particles in the flow behavior through models for the turbulent viscosity. The effect of turbulence on particle motion in gassolid suspension was analyzed by Yoshida and Masuda [18], Crowe [19], Alves et al. [17] and Zhang and Reese [20].

2. MATHEMATICAL MODEL

The equations used for the gas and solid phases, which are included in the description of the mathematical model, were developed with the Eulerian-Eulerian phenomenological approach. Differential equations in the balance are formulated for mass and momentum; treating the gas phase as incompressible. According to this approach, different phases can divide the same control volume at the same instant. The finite volume method numerically solved the set of governing conservation equations, mass and momentum, using the k-ε model on the gasphase and the kinematic viscosities analogy model on the solid-phase. The dense phase drag law was based on the Ergun equation with a modification of Gidaspow [2] and the dilute drag law was a modification of Wen &Yu [1], Yang et al. [21] and the continuous drag coefficient of Massarani [3]. Appropriate constitutive equations are specified for the description of the rheological physical properties of each phase and for the closure of the conservation equations.

2.1 Momentum and continuity conservation equations

The momentum and continuity conservation transient equations for each phase are as follows:

- Continuity equation, gas-phase

$$\frac{\partial}{\partial t}\left(\alpha_g \rho_g\right)+\frac{\partial}{\partial x}\left(\alpha_g \rho_g v_{g,x}\right)+$$
$$\frac{\partial}{\partial y}\left(\alpha_g \rho_g v_{g,y}\right)+\frac{\partial}{\partial z}\left(\alpha_g \rho_g v_{g,z}\right)=S_g^{\rho} \tag{1}$$

- Continuity equation, solid-phase

$$\frac{\partial}{\partial t}\left(\alpha_s \rho_s\right)+\frac{\partial}{\partial x}\left(\alpha_s \rho_s v_{s,x}\right)+\frac{\partial}{\partial y}\left(\alpha_s \rho_s v_{s,y}\right)+$$
$$\frac{\partial}{\partial z}\left(\alpha_s \rho_s v_{s,z}\right)=S_s^{\rho} \tag{2}$$

Where αg and αs are the volume fractions; ρg and ρs are the density of both phases; is the velocity vector, which can be decomposed into νx, νy, and νz, and Sρ represents the source term of continuity for each phase.

In the particular case of this research, the mass source terms are null due to the assumption of no mass transfer between phases.

$$S_g^{\rho}-S_s^{\rho}=0 \tag{3}$$

- Momentum equation, gas-phase (x direction)

$$\frac{\partial}{\partial t}\left(\alpha_g \rho_g v_{g,x}\right)+$$
$$\sum_{\xi=x,y,z}\frac{\partial}{\partial \xi}\left(\alpha_g \rho_g v_{g,x} v_{g,\xi}-\alpha_g \mu_g\left(\frac{\partial v_x}{\partial \xi}+\frac{\partial v_\xi}{\partial x}\right)\right)=S_{g,x}^{m} \tag{4}$$

$$S_{g,x}^{m}=\beta_{gs}^{m}\left(v_{s,x}-v_{g,x}\right)+\alpha_g \rho_g g_x-\frac{\partial p_g}{\partial x} \tag{5}$$

- Momentum equation, gas-phase (y direction)

$$\frac{\partial}{\partial t}\left(\alpha_g \rho_g v_{g,y}\right)+$$
$$\sum_{\xi=x,y,z} \frac{\partial}{\partial \xi}\left(\alpha_g \rho_g v_{g,y} v_{g,\xi} - \alpha_g \mu_g \left(\frac{\partial v_y}{\partial \xi} + \frac{\partial v_\xi}{\partial y}\right)\right) = S_{g,y}^m \tag{6}$$

$$S_{g,y}^m = \beta_{gs}^m \left(v_{s,y} - v_{g,y}\right) + \alpha_g \rho_g g_y - \frac{\partial p_g}{\partial y} \tag{7}$$

- Momentum equation, gas-phase (z direction)

$$\frac{\partial}{\partial t}\left(\alpha_g \rho_g v_{g,z}\right)+$$
$$\sum_{\xi=x,y,z} \frac{\partial}{\partial \xi}\left(\alpha_g \rho_g v_{g,z} v_{g,\xi} - \alpha_g \mu_g \left(\frac{\partial v_z}{\partial \xi} + \frac{\partial v_\xi}{\partial z}\right)\right) = S_{g,z}^m \tag{8}$$

$$S_{g,z}^m = \beta_{gs}^m \left(v_{s,z} - v_{g,z}\right) + \alpha_g \rho_g g_z - \frac{\partial p_g}{\partial z} \tag{9}$$

Where μg is the viscosity and Sgm represents the momentum transformation of the gas-phase. Analogous equations can be written for the solid-phase, where Sgm represents the momentum transformation in the solid-phase.

- Momentum equation, solid-phase (x direction)

$$\frac{\partial}{\partial t}\left(\alpha_s \rho_s v_{s,x}\right)+$$
$$\sum_{\xi=x,y,z} \frac{\partial}{\partial \xi}\left(\alpha_s \rho_s v_{s,x} v_{s,\xi} - \alpha_s \mu_s \left(\frac{\partial v_x}{\partial \xi} + \frac{\partial v_\xi}{\partial x}\right)\right) = S_{s,x}^m \tag{10}$$

$$S_{s,x}^m = \beta_{gs}^m \left(v_{g,x} - v_{s,x}\right) + \alpha_s \rho_s g_x - \frac{\partial p_s}{\partial z} \tag{11}$$

- Momentum equation, solid-phase (y direction)

$$\frac{\partial}{\partial t}\left(\alpha_s \rho_s v_{s,y}\right) + \sum_{\xi=x,y,z} \frac{\partial}{\partial \xi}\left(\alpha_s \rho_s v_{s,y} v_{s,\xi} - \alpha_s \mu_s \left(\frac{\partial v_y}{\partial \xi} + \frac{\partial v_\xi}{\partial y}\right)\right) = S^m_{s,y} \tag{13}$$

$$S^m_{s,y} = \beta^m_{gs}\left(v_{g,y} - v_{s,y}\right) + \alpha_s \rho_s g_y - \frac{\partial p_s}{\partial y} \tag{14}$$

- Momentum equation, solid-phase (z direction)

$$\frac{\partial}{\partial t}\left(\alpha_s \rho_s v_{s,z}\right) + \sum_{\xi=x,y,z} \frac{\partial}{\partial \xi}\left(\alpha_s \rho_s v_{s,z} v_{s,\xi} - \alpha_s \mu_s \left(\frac{\partial v_z}{\partial \xi} + \frac{\partial v_\xi}{\partial z}\right)\right) = S^m_{s,z} \tag{15}$$

$$S^m_{s,z} = \beta^m_{gs}\left(v_{g,z} - v_{s,z}\right) + \alpha_s \rho_s g_z - \frac{\partial p_s}{\partial z} \tag{16}$$

2.2 Turbulence

The k-ε turbulence model, in which the velocity and length scales are determined with separate transport equations, is used extensively for the determination of gas-phase turbulence because it offers good agreement between numeric effort and computational accuracy. Turbulent kinetic energy k is defined as the variation in velocity fluctuations. The edge turbulent dissipation ε is defined as the rate at which the velocity fluctuations dissipate and it has dimensions of k per time unit. The effective viscosity is the sum of dynamic viscosity and turbulent viscosity:

$$\mu_{g.ef} = \mu_g + \mu_{gt} \tag{17}$$

where μgt is the turbulence viscosity.

The model assumes that turbulent viscosity depends on the turbulent kinetic energy and the turbulent kinetic energy dissipation, through the relation proposed by Prandtl and Kolmogorov:

$$\mu_{g,t} = C_\mu \rho_g \frac{k^2}{\varepsilon} \tag{18}$$

Where Cμ is a constant of the turbulent model with a value of 0.09.

The values of k and ε are given directly by solution of the differential transport equations for the kinetic energy and turbulent dissipation rate (Equations (19) and (20), respectively):

$$\frac{\partial}{\partial t}\left(\alpha_g \rho_g k\right) + \nabla \cdot \left(\alpha_g \rho_g v_g k\right) = \nabla \cdot \left[\left(\mu_g + \frac{\mu_{g,t}}{\sigma_k}\right) \nabla k\right] + P_k - \alpha_g \rho_g \varepsilon \tag{19}$$

$$\frac{\partial}{\partial t}\left(\alpha_g \rho_g \varepsilon\right) + \nabla \cdot \left(\alpha_g \rho_g v_g \varepsilon\right) = \nabla \cdot \left[\left(\mu_g + \frac{\mu_{g,t}}{\sigma_\varepsilon}\right) \nabla \varepsilon\right] + \frac{\varepsilon}{k}\left(C_{\varepsilon 1} P_k - C_{\varepsilon 2} \alpha_g \rho_g \varepsilon\right) \tag{20}$$

Where Cϵ1, Cϵ2, σk and σϵ are constants of the k-ε turbulence model with values of 1.44, 1.92, 1.0 and 1.3, respectively.

Pk is the production of turbulence due to the viscous forces and buoyancy forces, expressed by

$$P_k = \mu_{g,t} \nabla u_g \cdot \left(\nabla v_g + \nabla v_g{}^T \right) -$$
$$\frac{2}{3} \nabla \cdot v_g \left(3\mu_{g,t} \nabla \cdot v_g + \alpha_g \rho_g k \right) + P_{kb} \tag{21}$$

The exact determination of the effective viscosity of the solid-phase is fundamental in attaining the radial distribution of the particles, and consequently, all the fluid dynamics variables [13]. In this research the dynamic viscosity for the solid-phase is null due to inviscid solids consideration.

The dispersed-phase zero-equation model consists in an analogy for the kinematic turbulent viscosities of the gas and solid phases; Equation (23) shows the expression for the solid phase:

$$v_{s,t} = \frac{v_{g,t}}{\omega} \tag{22}$$

Where ω is constant turbulent with a value of 10, relating the dispersed-phase kinematic turbulent viscosity (μs,t/ρs) to the gas-phase kinematic turbulent viscosity (μg,t/ρg).

2.3 Constitutive equations

2.3.1 Continuity between the phases:

$$\sum_i \alpha_i = \alpha_g + \alpha_s = 1 \tag{23}$$

Interphase momentum exchange: The coefficients of drag between fluid and particles are obtained from standard correlations with the negligence of acceleration. Without acceleration, friction on the wall or gravity, the unidimensional momentum balance for the gas-phase is

$$\alpha_g \frac{\partial p_g}{\partial x} - \beta_{gs}^{m}\left(v_g - v_s\right) = 0 \qquad (24)$$

The momentum transfer coefficient is obtained by comparison with Equation (24), which results in the Ergun equation [9]:

$$\frac{\Delta p}{\Delta x} = 150\frac{\alpha_s^2 \mu_g U}{\alpha_g^3\left(\theta_s d_p^2\right)} + 1.75\frac{U^2 \alpha_s \rho_g}{\alpha_g^3\left(\theta_s d_p\right)} \qquad (25)$$

Where U is the superficial velocity, U = αg (νg-νs) and θs is the particle sphericity; in this particular case, θs = 1 was used.

A comparison of Equations. (24) and (25) shows that for dense regimes (with αg< 0.8), the momentum transfer coefficient between gas phase and particles are in accordance with the following equation:

$$\beta_{gs}^{m} = 150\frac{\alpha_s^2 \mu_g}{\alpha_g d_p^2} + 1.75\frac{\left|v_s - v_g\right| \alpha_s \rho_g}{d_p} \qquad (26)$$

where dp is the particle diameter.

For porosities higher than 0.8, the expression for pressure drop results in the following equation for the interphase momentum transfer coefficient (Crowe [22]); this was also proposed by Gidaspow [2] and used in the simulation of Meier and Mori [13], Meier et al. [16], Alves et al. [14,17] and Bastos et al. [15] with a good agreement. In the Wen and Yu proposal the particles populational effect is desconsiderated, while Gidaspow and Ettehadieh (Gidaspow, [2]), the term fg -2.65 indicates the presence effect of the others particles in the fluid and actuates as a correction of usual Stokes law for the pressure drop of a simple particle.

$$\beta^{m}_{g,s} = \frac{3}{4} Cd \frac{\left|\vec{v}_s - \vec{v}_g\right| f_s \rho_g}{d_p} f_g^{-2.65} \tag{27}$$

The drag coefficient, cd, applied to Equation (27), is a function of the Reynolds number and behaves according to Equation (28), with a modified particle Reynolds number, and a power law correction, both functions of the continuous-phase volume fraction αg.

$$Cd = \alpha_g^{-1.65} \max\left[\frac{24}{\mathrm{Re}_p \alpha_g}\left(1 + 0.15(\mathrm{Re}_p \alpha_g)^{0.687}\right)\right] \tag{28}$$

For particle Reynolds numbers above 1,000 (turbulent flows), Equation (29) which is sufficiently large for inertial effects to dominate viscous effects (the inertial or Newton's regime), the drag coefficient becomes independent of Reynolds number:

$$\mathrm{Cd} = 0.44 \tag{29}$$

This uses the Wen and Yu correlation for low solid volume fractions αs< 0.2, and switches to Ergun's law for flow in a porous medium for larger solid volume fractions.

Note that this is discontinuous at the cross-over volume fraction.

In order to avoid subsequent numerical difficulties, it modifies the original Gidaspow model by linearly interpolating between the Wen and Yu and Ergun correlations over the range 0.7 <α_g> 0.8. The Reynolds equation for the particle is given by

$$\mathrm{Re_p} = \frac{\left|v_s - v_g\right| \rho_g d_p}{\mu_g} \tag{30}$$

Massarani (Massarani, [3]) proposed a modification in the drag coefficient correlation (Cd). This modification cover all values of

Reynolds for the particle, thus avoiding possible discontinuities in the calculation caused by the change of the flow regime due to axial velocity increase. Two new constants were introduced, particle sphericity functions, presented in Equation (31).

$$Cd = \left[\left(\frac{24}{K_1 Re_p} \right)^{0.85} + K_2^{\,0.85} \right]^{1.18} \tag{31}$$

where K1 and K2 are model constants and expressed by correlations, Equations 32 and 33:

$$K_1 = 0.843 \log_{10}\left(\frac{\theta_s}{0.065} \right) \tag{32}$$

$$K_2 = 5.31 - 4.88\theta_s \tag{33}$$

where θs is the particle sphericity.

2.4 Simulation

The simulated model consists basically of a turbulent jet flow, where two phases enter into contact. The gas-phase is formed by air and the solid-phase composed of catalysts. The particles are considered inviscid, smooth, spherical and inelastic, with a mean diameter of 50μm and a density of 2500kg/m3. Three different drag models were used in these cases (Wen and Yu [1], Gidaspow [2] and Massarani[3]) in order to evaluate which one better represents the jet flow dynamics. Due to gas-phase forces, which are responsible for the effective solid-phase distribution, the particles accelerate and move in the direction of the flow, characterizing the particle-laden jet regime in a short period of time. Measurements were taken at five axial levels of 120, 150, 180,210 e 240mm at the jet center, which has an internal diameter of 12mm, for Test and five cases with different initial conditions of velocity and solid mass loading rates, being all cases at its respective radial position in accordance with the data of Decker et al. [5].

2.5 Boundary conditions

At the entrance, all velocities and concentrations of both phases are specified. At the walls, the gas-phase velocity is zero and the solidphase velocity has a free-slip condition. The incompressible gas-phase pressure was defined at the exit, assuming atmospheric pressure. At the mesh boundaries, the opening condition was used. As inlet conditions for the Test case and other five cases of interest, the gas velocity (8 to 16m/s), the solid mass loading ratio (3 to 7) and the Reynolds number (5500 to 11000) are specified for each case and presented in Table 1.

2.6 Mesh and computational code

The non-structured mesh was composed of 170,000 control volumes. The details of the geometry and mesh are presented in Figure 1. The time step was on the order of 10-4s for the first second and on the order of 10-3s thereafter. Adaptation of the mathematical model for generation of the numerical model was achieved using the ANSYS CFX 12.0 commercial simulator, which is based on the finite volume method.

3. RESULTS AND DISCUSSION

3.1 Mesh tests

The radial profiles of the solid-phase mean velocity for the twophase jet flow were analyzed, using the proposed conditions by Decker et al. [5]. Eight meshes in two distinct configurations were tested, considering the jet nozzle into (Figure 2) and out the mesh, with different numbers of control volumes. Each pair consisted of approximately 100,000, 130,000, 170,000 and 200,000 control volumes.

Table 1. Inlet flow conditions.

Cases	U_0 (m/s)	η	Re
Test Case	11.13	4.5	7485
Case 1	8.25	6.34	5543
Case 2	9.64	5.41	6471
Case 3	11.78	4.26	7911
Case 4	13.23	3.69	8884
Case 5	15.83	3.09	10627

These meshes were tested and the one that was the best behavior for standard flow would be used to begin the simulations of the real process. In evaluating the dependence of flow on numerical mesh, it is known that the appropriate number of control volumes is of extremely important to avoid numerical errors, which is not possible with coarser meshes. The one-phase model was used for the tests, subjected to equations of continuity and momentum.

As a result, a mesh composed of 170,000 control volumes was chosen. This was shown to be in accordance with the established standards for both configurations used. Figure 3 shows these results for the Test case in both mesh configurations at z/D 12.5 and 15.0 without jet nozzle into the mesh (num*), jet nozzle into the mesh (num**) and experimental data, using Gidaspow drag correlation.

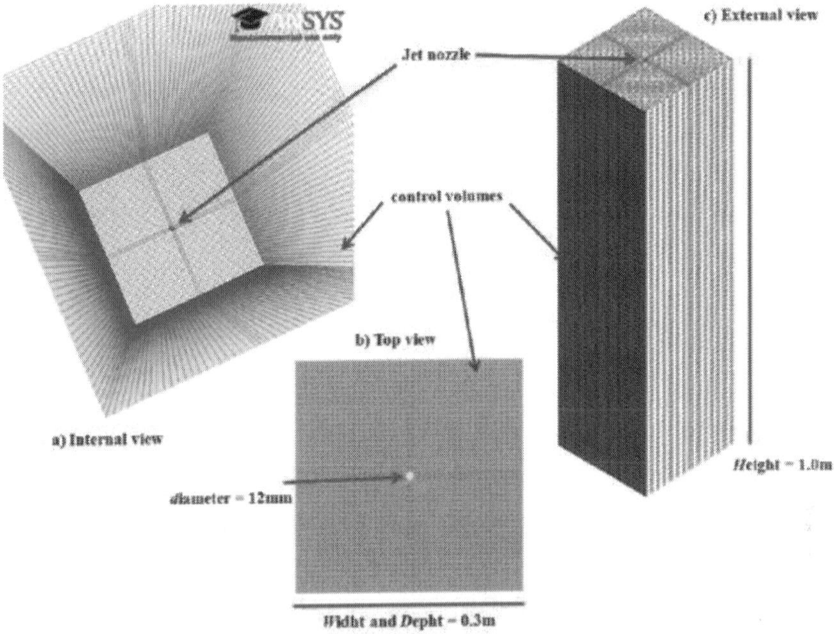

Figure 1. Geometry and Mesh details.

The simulations with different meshes resulted in the following observations:

In each one of the transversal sections, there are two distinct zones: a primary zone (where the particle velocities are high and quasi-constant), which extends from the radial center (r/R = 0) to approximately r = 0.007m, and a secondary zone (where the particle velocities are intermediary and decrease moderately), which is located at r = 0.007m to r = 0.02m (end of measurement line) for the two different configurations, in the two measured transversal sections [z/D = 12.5 and 15]. In the axial direction a single zone was observed, whereas for the same configuration in distinct sections overlap (Figure 3).

- Jet nozzle out the mesh

The velocity axial and radial profiles only reproduced the fluid dynamics behavior at the internal diameter of the jet nozzle (r = 0.006m), after this limit all results were underestimated compared to experimental results, due to not add to the calculations the effects of the jet outer boundaries in a direction to the jet core.

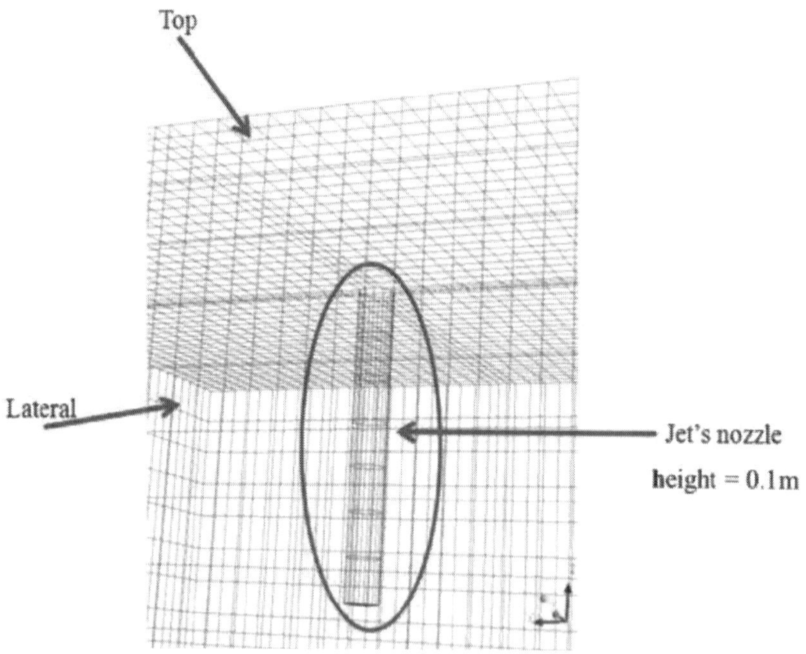

Figure 2. Jet's nozzle configurations.

- Jet nozzle into the mesh

The velocity axial and radial profiles reproduced completely the flow behavior as much quantitative as qualitatively.

- Axial dependency: There is not a significant influence on the mesh refinement in the axial direction. However, a slight variation was verified in the less refined mesh on the first and second cross sections (z/D 12.5 and 15.0).

- Radial dependency: In all cross sections the flow radial behavior was analogous compared to the experimental results for the Test case, according to Decker et al. [5], except for the coarser mesh.

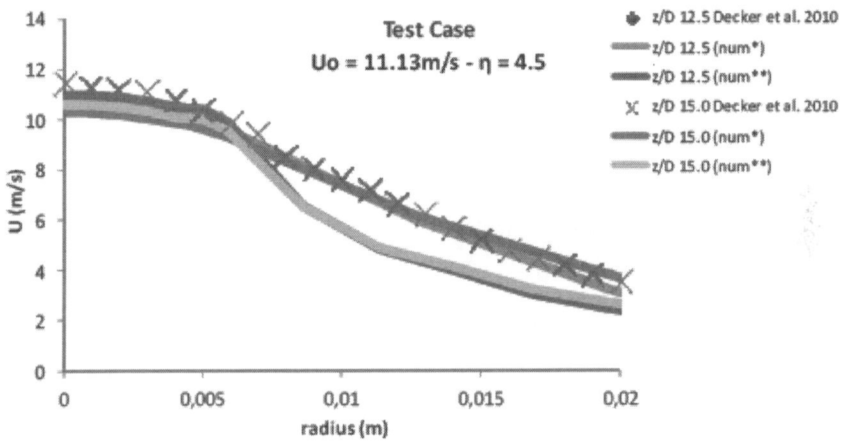

Figure 3. Radial profiles of the solid-phase mean velocity - Test case.

Figure 4 shows the respective measurement lines, in the center and line jet, and mainly the flow vectors, which indicates the influence of the effects of the jet outer boundaries in the direction to the jet center (jet into the geometry).

3.2 Dispersion model verification

Different simulations were accomplished to evaluate the flow dynamics in terms of the dispersion model used for the solid-phase. Among three drag correlations Wen & Yu symbolized by TCW, Gidaspow by TCG and Massarani by TCM.

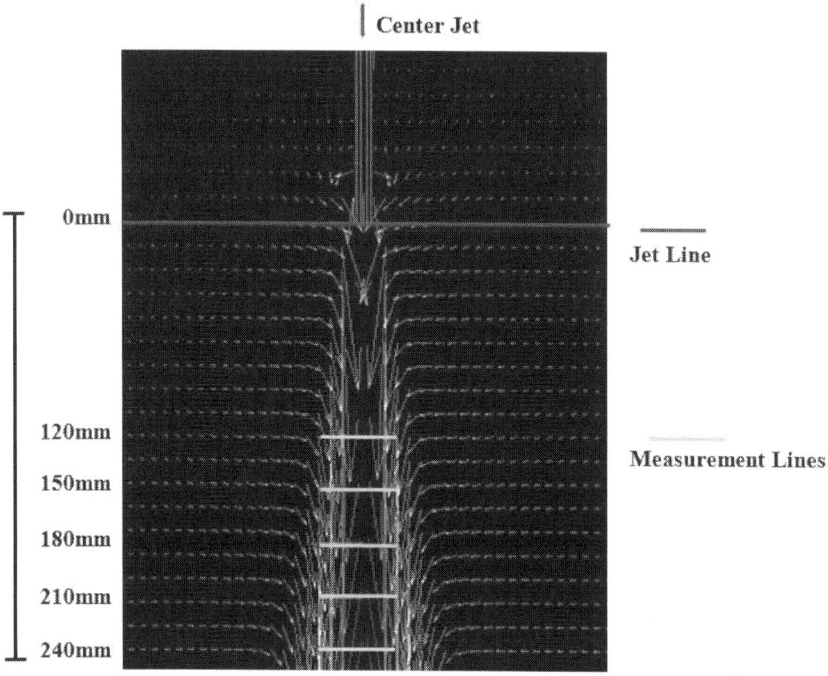

Figure 4. Lines and vectors with the jet into the mesh.

Figures 5a and 5b (close up), Figures 6a and 6b (close up)show the results of radial profiles of the solid-phase mean velocity for the Test case at z/D 12.5 and 15.0 compared to the experimental data of Decker et al. [5]. All the three models used showed good agreement with the experimental data. The drag models of Wen and Yu and Gidaspow showed similar results overlapped. However, the Massarani (TCM10) model showed a better tendency than them, what can be better observed in Figures 5b and 6b. Due to proximity of these results, it was chosen the last one to be closer to the experimental data and also the advantage of being continuous for the following five study cases.

3.3 Radial Profiles of the Solid-Phase Mean Velocity

The radial profiles of the solid-phase mean velocity were computed numerically at 7s of real time, using a complete mathematical model with Massarani drag correlation, k-ε turbulent model for the gas-phase, kinematic turbulent viscosities analogy for the solid-phase and the time averaging procedure at each time step.

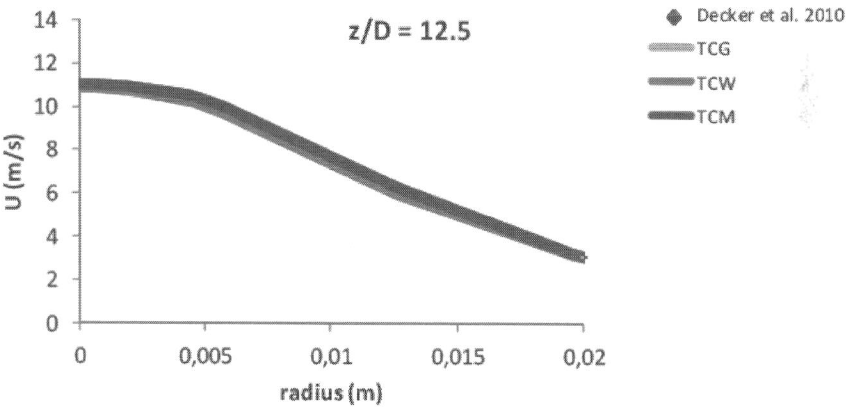

Figure 5a. Radial profile of the solid-phase mean velocity at z/D12.5 – drag correlation analysis

Figures 7 and 8 shows the obtained simulation results for cases 1, 2, 3, 4 and 5 at the closer cross sections z/D 12.0 and 12.5, respectively (Figure 4). Although the simulated results under predicted the solidphase mean velocity in the near the jet boundaries, one can see that the profiles in Figure 7 show the tendency of the experimental data. In Figure 8 this can only be observed for case 5, however more attenuated. Cases 1,2,3 and 4 showed good agreement between numerical and experimental data.

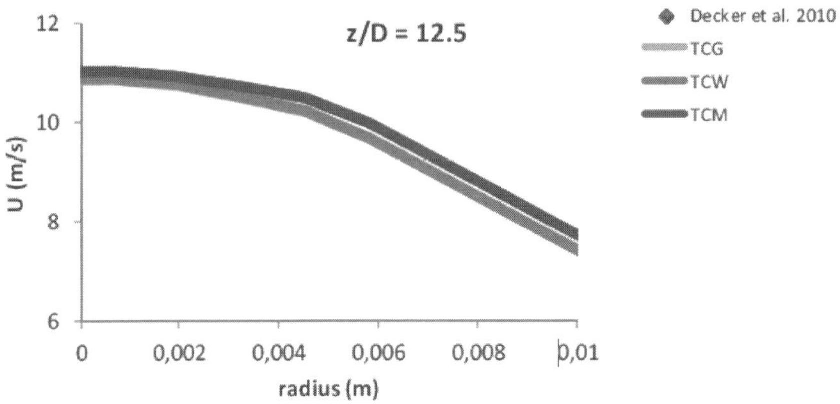

Figure 5b: Radial profile of the solid-phase mean velocity at $z/D12.5$ – closeup.

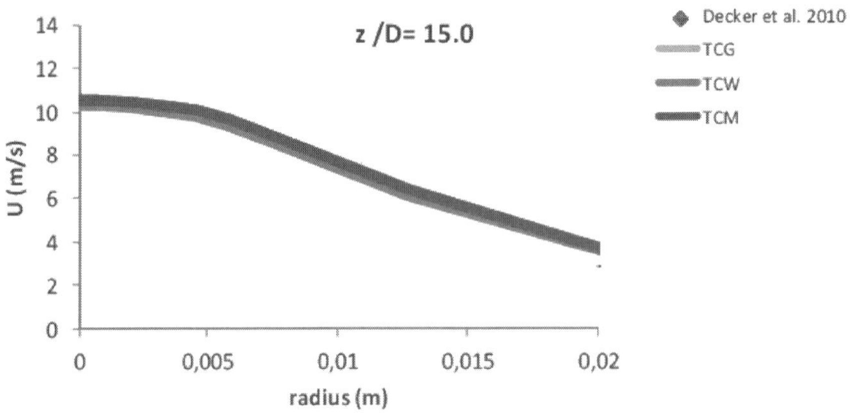

Figure 6a: Radial profile of the solid-phase mean velocity at $z/D15.0$ – drag correlation analysis.

It is possible to observe the same behaviors, both quantitative and qualitative, at different radial and axial positions. The solid-phase velocity tends to remain similar in all planes and in all cross-sections near the central jet. However, this velocity, whether near or far from jet limit decreases axially and radially. It can be observed in each plane there are variations in that solid-phase velocity at the boundary of the jet, especially at the higher velocities (cases 4 and 5) and in the end of measurement line.

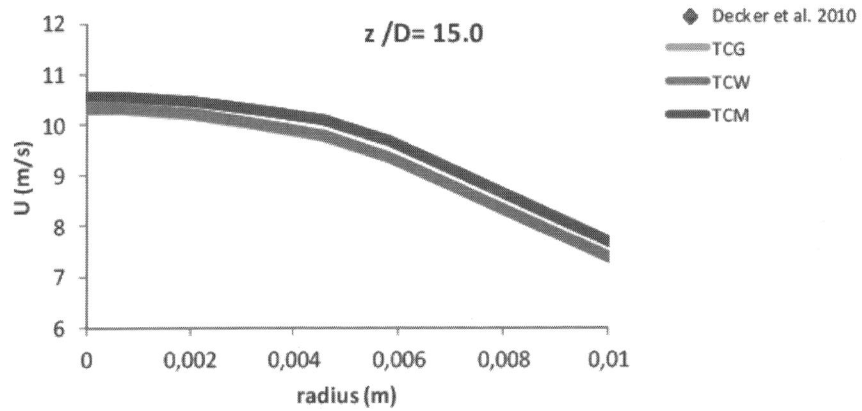

Figure 6b: Radial profile of the solid-phase mean velocity at $z/D15.0$ – closeup.

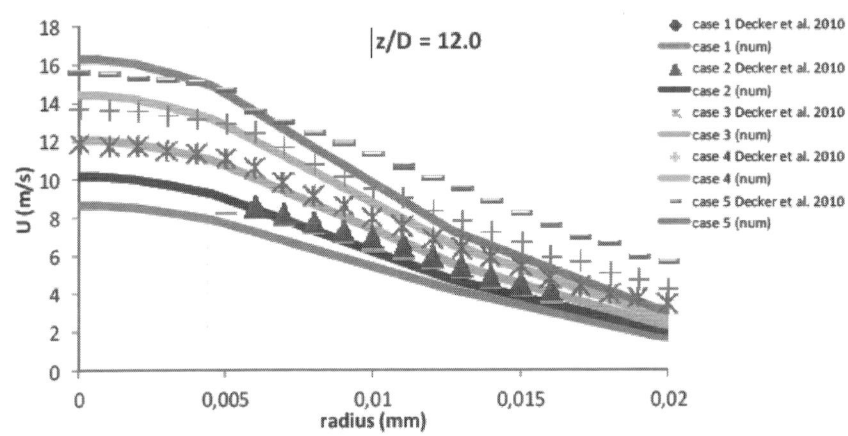

Figure 7: Radial profile of the solid-phase mean velocity at z/D12.0.

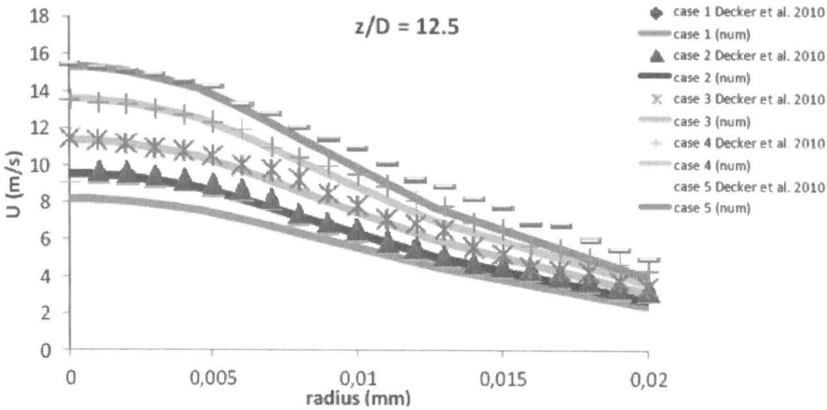

Figure 8. Radial profile of the solid-phase mean velocity at z/D12.5.

The simulated results for the other three cross sections (z/D 15.0, 17.5 and 20.0) for five interest study cases presented complete agreement between numerical and experimental data. Figures 9, 10 and 11 show these results. It is possible to observe the same behaviors, both quantitative and qualitative, at different radial and axial positions. The solid-phase velocity tends to remain similar in all cross-sections in the center jet and near the jet boundaries. It can be observed in all crosssections investigated, as in Test case, two distinct zones that extend at same radial points (0 to approximately 0.007 and 0.007 to 0.02m). Also, in the axial direction a single zone was observed. The solid-phase velocity was for the five cases, in fact, higher in the primary zone than in the second zone at all axial positions, in agreement with experimental data (Decker et al. [5]) and as seen in Yan et al. [23].

4. CONCLUSIONS

The objectives of this research was analyzing the influence of the different mathematical models of drag for the gas-solid jet flow prediction and confirm the turbulence models for each phase, in special the kinematic turbulent viscosities analogy for jet flow. Numerical radial profiles of the solid-phase mean velocity were analyzed for all experimental measured positions and the results were compared. The simulation results obtained regarding drag and solid-phase dispersion correlations dependence on the three different drag correlation used in this work were analyzed. The predicted radial profiles of the solid-phase mean velocity showed excellent agreement with the measurements in the flow cross sections for the Test case and five study cases in a jet.

ACKNOWLEDGEMENTS

The financial support received from the PETROBRAS is gratefully acknowledged by the authors.

REFERENCES

1. Wen CY, Yu YH (1966) Mechanics of Fluidization. Chem Eng Progr Symp 62: 100-111.
2. Gidaspow D (1994) Multiphase flow and fluidization: continuum and kinetic theory descriptions. Academic Press, San Diego, California, USA.
3. Massarani G (2002) Fluidodinâmica em Sistemas Particulados. (2ndedn), E-Papers Serviços Editorials.
4. Fairweather M, Hurn JP (2008) Validation of an anisotropic model of turbulent flows containing dispersed solid particles applied to gas-solid jets. Comput Chem Eng 32: 590-599.
5. Decker RK, Meier HF, Fritsching U, Mori M (2010) Analysis of Particle Interaction with Coherent Structures in a Two-Phase Mixing Jet. Mult Scien Techn 22: 1-30.

6. Hardalupas Y, Taylor AMK, Whitelaw JH (1989) Velocity and particle-flux characteristics of turbulent particle-laden jets. Proc R Soc Lond A 426: 31-78.

7. Hadinoto K, Jones EN, Yurteri C, Curtis JS (2005) Reynolds number dependence of gas-phase turbulence in gas–particle flows. Int J Multiphase Flow 31: 416-434.

8. He Y, Rudolph V (1995) Gas-solids flow in the riser of a circulating fluidized bed. Chem Eng Sci 50: 3443-3453.

9. Theologos KN, Markatos NC (1993) Advanced modeling of fluid catalytic cracking riser-type reactors. A I Ch E Journal 39: 1007–1017.

10. Ali H, Rohani S (1997) Dynamic modeling and simulation of a riser-type fluid catalytic cracking unit. Chem Eng Technol 20: 118–130.

11. Soo SL (1967) Fluid dynamics of multiphase systems. Blaisdell Publishing co, Waltham, Massachusetts, USA.

12. Rietema K, van den Akker HEA (1983) On the momentum equations in dispersed two-phase systems. Int J Multiphas Flow 9: 21-36.

13. Meier HF, Mori M (1998) The Eulerian-Eulerian-Lagragian Model for Cyclone Simulation, Proceedings of the Fourth European Computational Fluid Dynamics Conference.

14. Alves JJN, Mori M (1998) Fluid dynamic modelling and simulation of circulating fluidized bed reactors: analyses of particle phase stress models. Comput Chem Eng 22: S763-S766.

15. Bastos JCSC, Rosa LM, Mori M, Marini F, Martignoni WP (2008) Modeling and Simulation of a Gas-Solid Dispersion Flow in a High-Flux Circulating Fluidized Bed (HFCFB) Riser. Catal Today 130: 462-470.

16. Meier HF, Mori M (1999) Anisotropic Behavior of the Reynolds Stress in Gas-Solid Flows in Cyclones. Powder Technol 101: 108-119.

17. Alves JJN, Meier HF, Martignoni WP, Mori M (2001) The Effect of Turbulence on the Flow Pattern of Circulating Fluidized Bed Reactors. Proceedings of ENPROMER 3: 1351-1356.

18. Yoshida H, Masuda H (1980) Model Simulation of Particle Motion in Turbulent Gas-Solid Pipe Flow. Powder Technol 26: 217-220.

19. Crowe CT (2000) On Models for Turbulence Modulation in Fluid-Particle Flows. Int J Multiphase Flow 26: 719-727.

20. Zhang Y, Reese JM (2003) Gas turbulence modulation in a two-fluid model for gas–solid flows. A I Ch E Journal 49: 3048–3065.

21. Yang N, Wang W, Ge W, Wang L, Li J (2004) Simulation of heterogeneous structure in a circulating fluidized-bed riser by combining the two-fluid model with the EMMS approach. Ind Eng Chem Res 43: 5548-5561.

22. Crowe CT (1982) Review - Numerical Models for Dilute Gas-Particle Flows. J Fluids Eng 104: 297–303.

23. Yan J, Luo K, Fan J, Tsuji Y, Cen K (2008) Direct Numerical Simulation of Particle Dispersion in a Turbulent Jet Considering Interparticle Collisions. Int J Multiphase Flow 34: 723-733.

2

NUMERICAL SIMULATIONS OF TWO-PHASE FLOW IN A SELF-AERATED FLOTATION MACHINE AND KINETICS MODELING

Hassan Fayed [1] and Saad Ragab [2,*]

[1]Numerical Porous Media Center, King Abdullah University of Science and Technology (KAUST), Thuwal 23955-6900, Saudi Arabia

[2]Department of Engineering Science and Mechanics, Virginia Tech, Blacksburg, VA 24061, USA

ABSTRACT

A new boundary condition treatment has been devised for two-phase flow numerical simulations in a self-aerated minerals flotation machine and applied to a Wemco 0.8 m^3 pilot cell. Airflow rate is not specified *a priori* but is predicted by the simulations as well as power consumption. Time-dependent simulations of two-phase flow in flotation machines are essential to understanding flow behavior and physics in self-aerated machines such as the Wemco machines. In this paper, simulations have been conducted for three different uniform bubble sizes (d_b = 0.5, 0.7 and 1.0 mm) to study the effects of bubble size on air holdup and hydrodynamics in Wemco pilot cells. Moreover, a computational fluid dynamics (CFD)-based flotation model has been developed to predict the pulp recovery rate of minerals from a flotation cell for different bubble sizes, different particle sizes and particle size distribution. The model uses a first-order rate equation, where models for probabilities of collision, adhesion and stabilization and collisions frequency estimated by Zaitchik-2010 model are used for the calculation of rate constant. Spatial distributions of dissipation rate and air volume fraction (also called void fraction) determined by the two-phase simulations are the input for the flotation kinetics model. The average pulp recovery rate has been calculated locally for different uniform bubble and particle diameters. The CFD-based flotation kinetics model is also used to predict pulp recovery rate in the presence of particle size distribution. Particle number density pdf and the data generated for single particle size are used to compute the recovery rate for a specific mean particle diameter. Our computational model gives a figure of merit for the recovery rate of a flotation machine, and as such can be used to assess incremental design improvements as well as design of new machines.

KEYWORDS

minerals flotation machines; two phase flows; flotation kinetics; rate constant; particles size distribution

1. INTRODUCTION

Mineral flotation machines are classified into two main types; forced air and self-aerated machines. Wemco machines are widely used self-aerated machines where no air pumping mechanism is required, which simplifies flotation plant design and operation. The rate of air flow and power consumption of Wemco machines depend on the flow structure and the hydrodynamics within the pulp volume. For a given machine, rate of airflow depends on the rotor speed (RPM) among other operating conditions such as rotor blades and disperser design. Instantaneous airflow rate in a Wemco machine is not known *a priori* and depends on machine design and operating conditions. Therefore, computational fluid dynamics (CFD) simulation of a Wemco machine is a challenging problem because airflow rate cannot be specified but it has to be an outcome of the simulation. Moreover, rate of air flow may vary significantly with time to the extent that air is temporarily "exhaled" by the standpipe instead of being "inhaled". Computer simulations of such a machine should predict the time-history of the rate of air flow, and the average rate is an output. The unknown rate of air flow and the possibility of "breathing" require careful treatments of the standpipe and pulp–froth interface boundary conditions. Koh and Schwarz [1] conducted CFD simulation of the self-aerated flotation machine Denver-Metso Minerals. In that machine, air flow rate depends on suction pressure created by the impeller, the hydrostatic head of the pulp, and the frictional losses along the delivery shaft from the inlet valve to the impeller. The air motion was not simulated in the standpipe. They predicted the air flow rate iteratively during the simulation by applying pressure loss formula to find the pressure drop in the standpipe. An empirical constant in the formula was adjusted for CFD simulations to match the experimental data.

Computational domains of flotation machines are large, and flow physics are complex involving multi-phase flow turbulence. Even two-phase flow simulations of flotation machines are time consuming and require large computational resources. Some approaches have been used to reduce computational costs for two-phase flow; see, for example, the approach by Tiitinen *et al.* [2], where sector based simulations were used to reduce the number of grid nodes. Bubble size is one of the most important parameters that affect the air holdup of the pulp phase. A spectrum of bubble sizes exists in flotation machines depending on air flow rate and turbulence parameters. To predict such bubble size distribution, another set of equations that describes a population balance can be solved in the course of CFD simulation (Kerdouss *et al.* [3]). This approach increases the computational demands where transport equation for each size group has to be implemented. A more feasible approach is to conduct a parametric study for different uniform bubble sizes to study their effects on air holdup and rate constant.

One of the main characteristics of mechanical flotation machines is to agitate the slurry and disperse air bubbles throughout the pulp volume. In order to assess the performance of flotation machines, it is important to know the spatial distribution of dispersed bubbles within the tank which directly affects air hold up and rate constant. The current CFD simulations are parametric study of two-phase flow in Wemco 0.8 m^3 that provide the hydrodynamic data and air volume fraction spatial distribution for uniform bubble size in the pulp phase. Three different bubble sizes—d_b = 0.5, 0.7 and 1.0 mm—are used to investigate the effects of the bubble size on air flow rate, air holdup and rate constant. The paper is organized as follow. Section 2 presents the governing equations for two-phase flow. Machine geometry is presented in Section 3, and simulation results are discussed in Section 4. Flotation model and results are presented and discussed in Section 5, and conclusions are summarized in Section 6.

2. EULER-EULER TWO-FLUID MODEL

A practical approach for two-phase simulations is the Euler-Euler approach in which both phases are modeled by volume-averaged equations [4]. The motion of the two continuous phases is described by the unsteady Reynolds-averaged Navier–Stokes (RANS) equation:

Continuity equation:

$$\frac{\partial(\alpha_i \rho_i)}{\partial t} + \nabla.\left(\alpha_i \rho_i \vec{V}_i\right) = 0 \tag{1}$$

Momentum equation:

$$\frac{\partial(\alpha_i \rho_i \vec{V}_i)}{\partial t} + \nabla.(\alpha_i \rho_i \vec{V}_i \vec{V}_i) = -\alpha_i \nabla P + \nabla.[\mu_{i,\text{eff}} (\nabla \vec{V}_i + (\nabla \vec{V}_i)^T)] + S_i + M_i \qquad (2)$$

where (i = 1) denotes water phase and (i = 2) denotes gas phase, ∇P is the modified pressure to include the gravity effects, S_i describes any external momentum source and Mi is the interfacial force that acts on phase (i) due to the presence of other phases. In the present simulations, momentum exchanges between the two phases due to drag and buoyancy on bubbles are the only mechanisms that couple the motion of the two phases. Bubbles are deformable fluid particles when moving in high shear rate regions such as in minerals flotation machines. Schiller-Naumann drag model [5] has been used to estimate drag coefficient of air bubbles. Effects of bubbles deformations on the values of drag coefficient are neglected in Schiller-Naumann drag model [5].

Shear stress transport (SST) turbulence model has been used to model turbulence transport where two transport equations are solved. There is no such universal turbulence model for two-phase flow, particularly at high volume fraction [6]. Reynolds stress model is the most adequate model for swirling flows. However, Reynolds stress model (RSM) closes the RANS equations by solving six additional transport equations for averaged Reynolds stress terms, and that require large computational resources. Geometry and flow physics of minerals flotation machines are large and complex. Therefore, using RSM turbulence model is not feasible for such application and SST model has been used in our CFD study.

3. CELL GEOMETRY AND SIMULATIONS PARAMETERS

The main components of Wemco-0.8 m³ machine (FLSmidth, Salt Lake City, UT, USA) as shown in Figure 1include a six-blade rotor, a disperser, a draft tube, and a standpipe. Details of the different components are shown in Figure 2. The disperser has 34 holes arranged in two parallel rows. Seventeen semi-circular rods are attached to the inner surface of the disperser. Air is drawn into the machine through a hole in the top of standpipe. The machine is assembled in Figure 3. The simulated Wemco-0.8 m^3 model does not have a disperser hood or tank baffles.

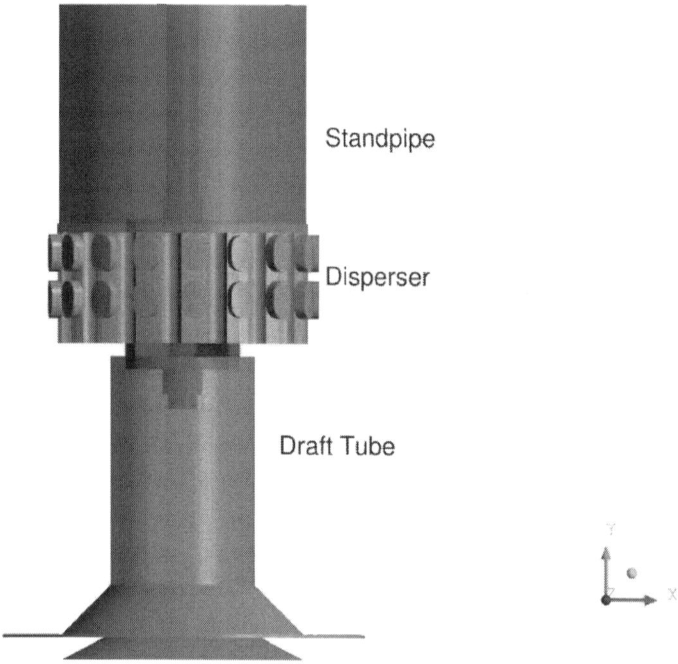

Figure 1. Core of the Wemco-0.8m^3 machine: rotor, disperser, draft tube, and standpipe.

The boundary condition treatment at the air opening in top of the standpipe allows air to enter or exit from the standpipe above the rotor. There is no valve or any other device that obstructs air from flowing back to the ambient through the air inlet hole. If airflow reverses direction it may also carry a small amount of water with it. A container, which is connected to the standpipe only through the air inlet hole, is added. Its function is to retain any water that might be expelled and is to be retrieved when air flows back into the machine.

Atmospheric pressure is prescribed at the top of that container. This treatment is unique to the present simulations of a self-aerated machine, and it allows the machine to "breath" for some operating conditions. An important operating assumption used in the present simulations is that the amount of water within the tank remains constant during operation. That is to say no water is allowed to flow over the weir. This condition may not be precisely satisfied in the actual machine operation; but it is a realizable situation. To guarantee that no water exits the tank, the tank wall is extended vertically above the weir edge thereby creating an overflow tank. The overflow tank is a mere vertical extension of the actual tank walls. Atmospheric pressure is prescribed at the top of the overflow tank. The actual tank is initialized with 100% water up to a certain level to be defined later, and the rest of the actual tank and overflow tank are initialized with 100% air.

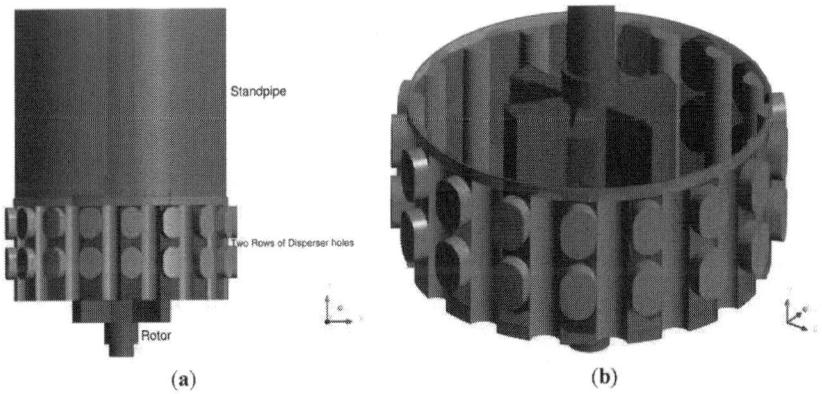

Figure 2. Details of Wemco-0.8m^3 rotor, disperser, and standpipe; (**a**) relative vertical position of rotor and disperse; (**b**) rotor and disperser.

The function of the overflow tank is to permit the water level to rise due to the accumulation of air in the pulp, and at the same time it does not allow water to exit the computational domain. No boundary conditions are needed at the interface between the actual tank and the overflow tank. The governing equations of the two-phase flow are solved in both tanks allowing air and water to flow back and forth between the tanks as required by the transport equations. The part of the computational domain below the initial water level is initialized (t = 0) with 100% water and the part above that level is initialized by 100% air. A multi block structured grid has been generated for the current simulations using ICEM CFD grid generator. The total number of nodes is around 2.4×10^6. The size of the cells and number of nodes on rotor, disperser walls among other parts have been chosen to resolve boundary layer flow on these walls. Two phase flow simulations have been conducted using ANSYS CFX 12.0 commercial CFD software (ANSYS, Inc., Canonsburg, PA, USA). Governing equations are discretized in space using second order advection scheme and solved in time by using second order backward Euler method. The unsteady calculations have been performed on 10 parallel processors. The time step (Δt) is equal to 0.0007 s, at 620 rpm impeller speed that allows impeller volume to rotate ($\Delta\theta$ = 1.5) degrees per time step.

4. TWO PHASE FLOW RESULTS

4.1. Air Flow Rate and Power

The rate of air flow drawn into the machine varies with time as shown in Figure 4a for three assumed bubble diameters. It is evident that the airflow is unsteady and momentarily may reach zero or become slightly negative; air is flowing out of the standpipe. Little effects of the bubble diameter are observed. The power (the computed rotor torque is multiplied by angular velocity) as depicted in Figure 4b also varies with time; it also shows negligible effects of bubble diameter. Air flow rate fluctuates around 0.3 m^3/min and power consumption fluctuate around 1.4 kW. Also, Figure 4a shows that simulations reach steady state after 8 s. The running average of air flow rate and power are defined by:

$$\tilde{Q}_a(t) = \frac{1}{t}\int_0^t Q_a(\tau)\,\mathrm{d}\tau \tag{3}$$

$$\tilde{P}(t) = \frac{1}{t}\int_0^t P(\tau)\,\mathrm{d}\tau \tag{4}$$

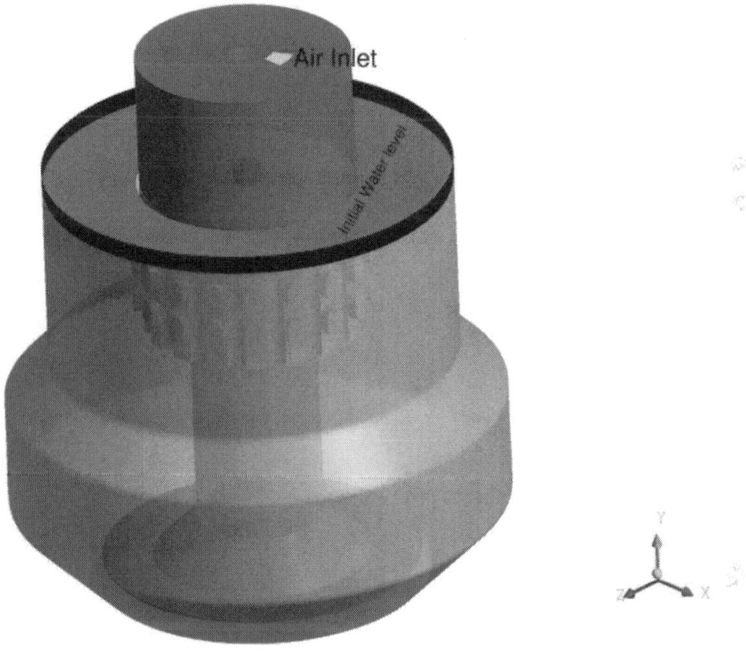

Figure 3. Wemco-0.8 m³ assembly: tank, air-inlet and initial water level.

Calculations of the running average include initial transient variations of air flow and power. As shown in Figure 5, the time running averaged air flow rate is converging to 0.35 m^3/min and the power to 1.43 kW. The time-averaged air capacity coefficient $C_a = Q_a/ND^3$ is 0.053 for Wemco 0.8 m^3. Nelson *et al.* [7] reported measured values of C_a for large Wemco cells (for example, Wemco-

160 m^3 SmartCell) in the range of 0.13–0.17. Nelson's *et al.* [7] work is the only available experimental study for Wemco machines that reports those values. It is notable that the experimental values are much greater than the present CFD prediction for the smaller pilot Wemco 0.8 m^3 model. Values for capacity coefficient are machine dependent, and this discrepancy is due to the differences in machine design as well as operating conditions. In flotation process, air rise velocity in the tank is one of the important design parameters. It is used to determine tank diameter. Minerals flotation engineers usually consider air rise velocity to be around 1.0 m/min for efficient flotation process. This value does not depend on machine type or size. In the current CFD study, the predicted air rise velocity (Q_a/A_c) equals 0.92 m/min which is close to the recommended value. Moreover, predicted air rise velocity lies in the measured range of 0.80–1.24 m/min as reported by Nelson *et al.* for Wemco-160 m^3.

4.2. Flow Pattern and Velocity Field

The flow pattern in the Wemco rotor is very complex and unsteady. Analysis of the velocity field and air volume fraction in the rotor region will shed light on the principle of the Wemco machine operation. Water superficial radial velocity contours on horizontal planes that cut the rotor are depicted in Figure 6a–f at time t = 14.35 s. In Figure 6c, we see that the two blades at positions twelve-o'clock and six-o'clock have the strongest radial outflow, whereas the two blades in the positions four-o'clock and ten-o'clock have inward radial flow. The radial inward flow is entrained by the rotor from the secondary flow below the disperser. The pumping situation switches at higher elevations as shown in Figure 6e. Here, the radial flow at the blades at twelve-o'clock and six-o'clock is blocked whereas that at the blades at four-o'clock and ten-o'clock is strong and outward. Hence, water exits the rotor non-uniformly both circumferentially and vertically. Lack of flow periodicity in the Wemco rotor has also been predicted by CFD simulations of single-phase flow in Wemco 250 and 300 conducted by the present authors.

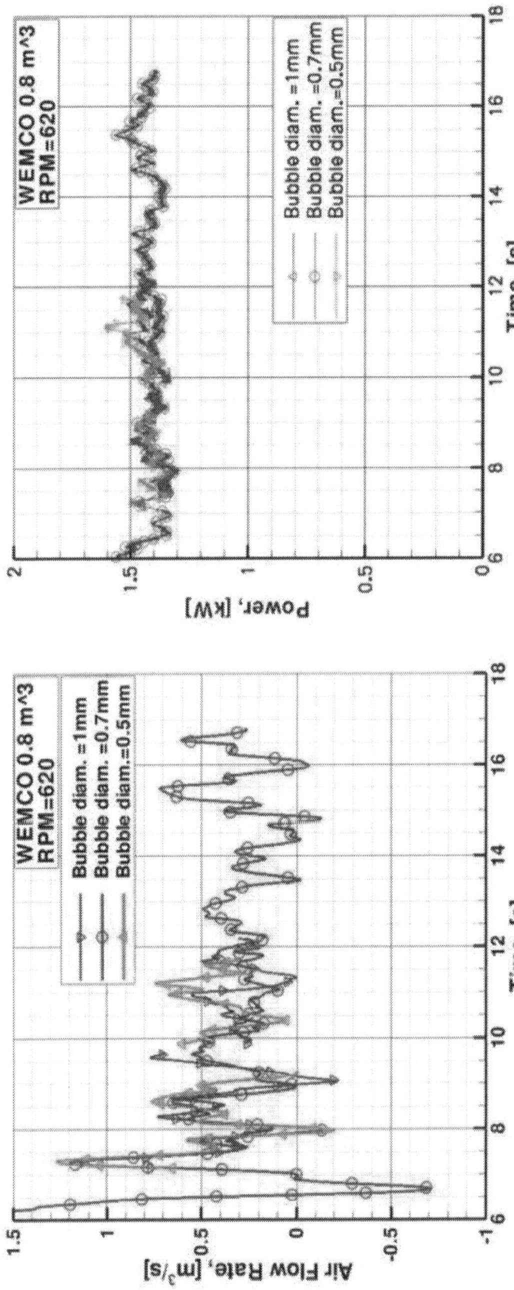

Figure 4. Effects of assumed bubble diameter on airflow rate and power, d_b = 0.5, 0.7 and 1.0 mm; (**a**) Variation of air volume flow rate with time; (**b**) Variation of power with time.

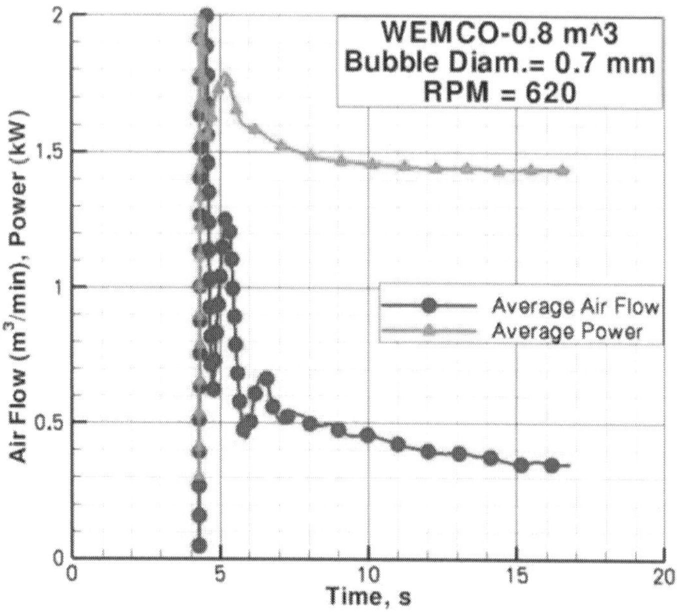

Figure 5. Running average of air flow rate and power.

The velocity contours in the same planes are shown in Figure 7a–f at t = 15.4 s. At this instant, the maximum airflow has been observed and the water content in the rotor is a maximum. The important qualitative difference in the flow at this instant of time and the earlier one is the change in the direction of the vertical velocity at the rotor top. Water flows from the rotor into the standpipe mainly through the gap between rotor and disperser, and then falls down back from the standpipe mixed with air on rotor blades. Then rotor blades pump this mixture through the upper row of disperser holes to the tank. Figure 8a,b show the air volume fraction in a vertical mid plane that passes through the axis of the machine at two different times where the airflow rate is minimum and maximum, respectively. A vortex is formed in the standpipe, and water adheres to the standpipe walls under the effects of centrifugal forces (*i.e.*, water flows up into the standpipe with high circumferential velocity component). Consequently, water head in the standpipe increases continuously until it reaches its maximum level. At this instant, water breaks away from the standpipe walls and falls under gravity to the rotor tips capturing air with it as a mixture as depicted in Figure 8a,b. Water velocity vectors in a mid-plane that passes through the axis of the machine at two different time levels are shown in Figure 9a,b. Figure 9a shows the water velocity vectors at the maximum airflow instant. Water is discharged radially through the upper holes of the disperser where some water falls down on the rotor tips. Water velocity vectors at minimum airflow rate instant are shown in Figure 9b. Secondary flow below the disperser has been noticed for both time instants.

Three uniform bubble diameters of 0.5, 0.7, and 1.0 mm have been used to study the effects of bubble size on air holdup. Air holdup is computed by integration of (local) air void fraction over the pulp volume and then divided by pulp volume. The computed air holdup for 0.5, 0.7, and 1.0 mm bubble diameters are 9.3%, 7.4%, and 5.7%, respectively. It is observed that air holdup increases as bubbles size decrease. In this study, drag and buoyancy are the only considered interfacial forces between air bubbles and the liquid phase. Drag on air bubbles depends on bubble cross sectional area (*i.e.*, d_b^2) while buoyancy depends on the bubble volume (*i.e.*, d_b^3). Therefore, drag is the dominant force for small bubbles while buoyancy is the dominant force for large bubbles. Small bubbles have negligible effective inertia and experience lower slip velocity to the liquid phase. They tend to penetrate deeper into the tank and take longer time (*i.e.*, retention time) to escape out at the free surface of the pulp, which results in higher air holdup. Large bubbles experience higher slip velocity and escape quickly from the tank because of the larger buoyancy that results in lower air holdup.

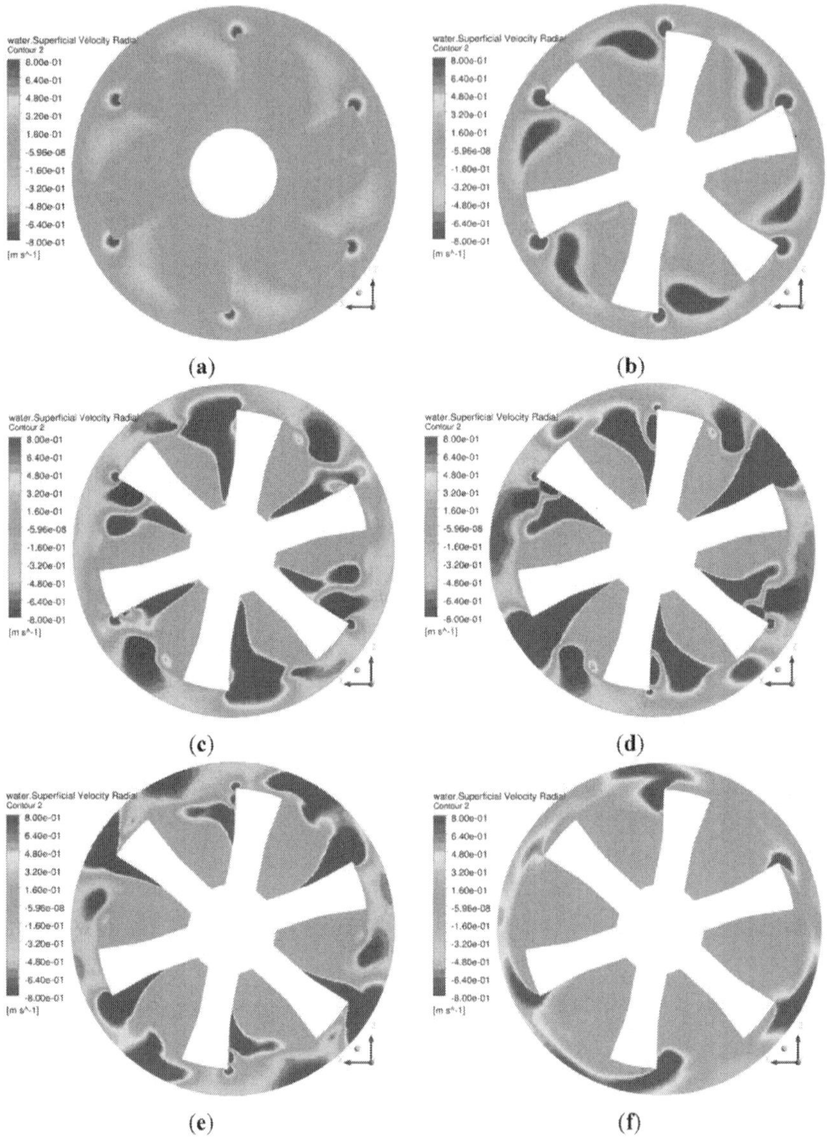

Figure 6. Water superficial radial velocity contours in horizontal planes cutting rotor blades, t = 14.35 s, d_b = 0.7 mm; (**a**) y = 0.86 m; (**b**) y = 0.87 m; (**c**)y = 0.9 m; (**d**) y = 0.92 m; (**e**) y = 0.96 m; (**f**) y = 1.10 m.

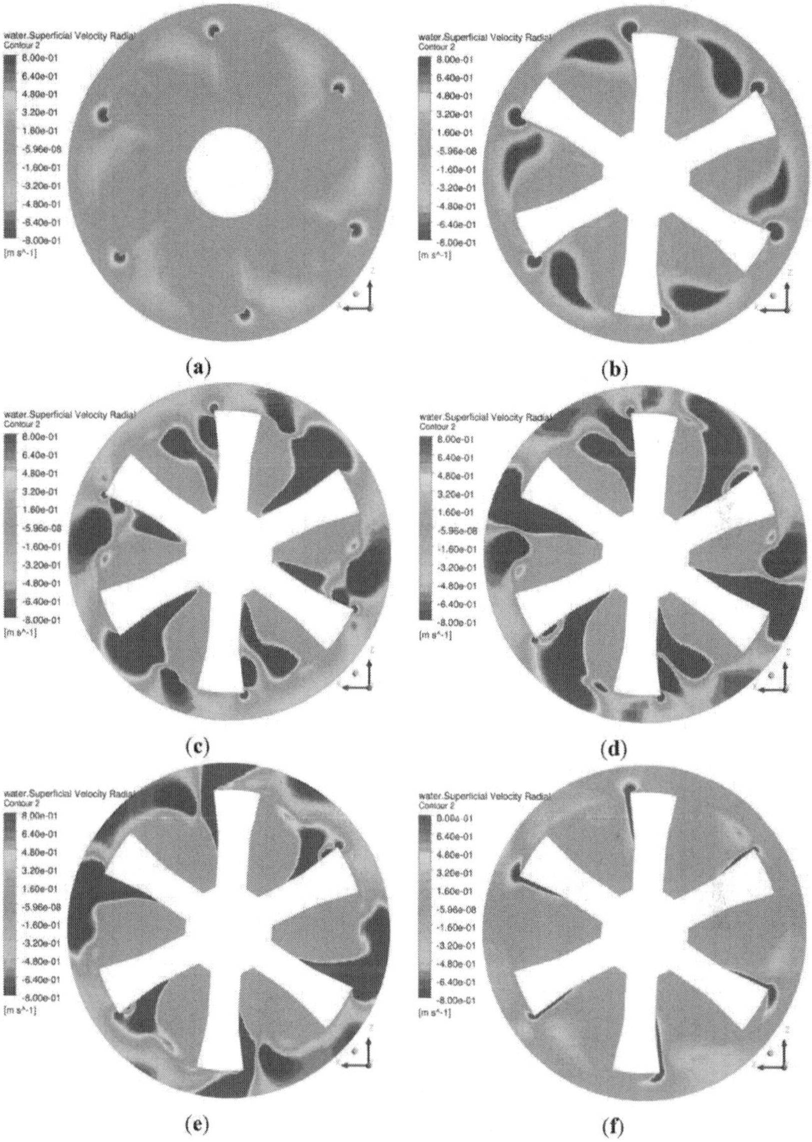

Figure 7. Water superficial radial velocity contours in horizontal planes cutting rotor blades, t = 15.4 s, d_b = 0.7 mm; (**a**) y = 0.86 m; (**b**) y = 0.87 m; (**c**) y= 0.9 m; (**d**) y = 0.92 m; (**e**) y = 0.96 m; (**f**) y = 1.10 m.

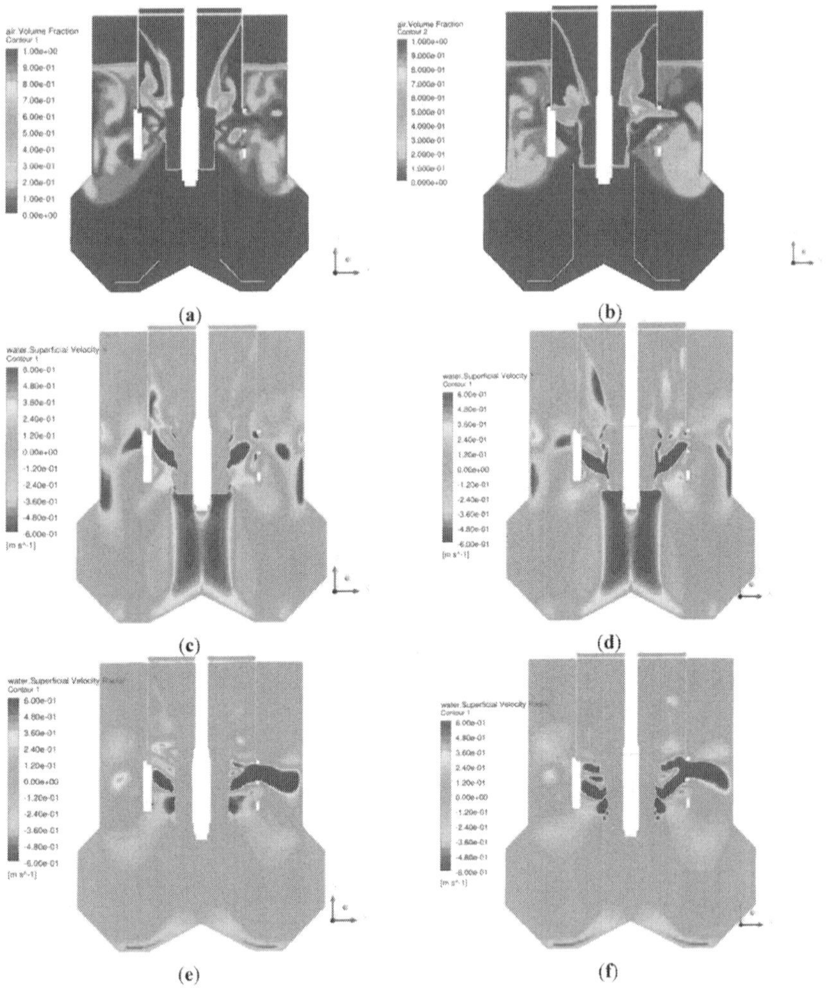

Figure 8. Air volume fraction and velocity contours for, d_b = 0.7 mm (**a**) air volume fraction, t = 14.35 s; (**b**) air volume fraction, t = 15.4 s; (**c**) water superficial vertical velocity, t = 14.35 s; (**d**) water superficial vertical velocity,t = 15.4 s; (**e**) water superficial radial velocity, t = 14.35 s; (**f**) water superficial radial velocity, t = 15.4 s.

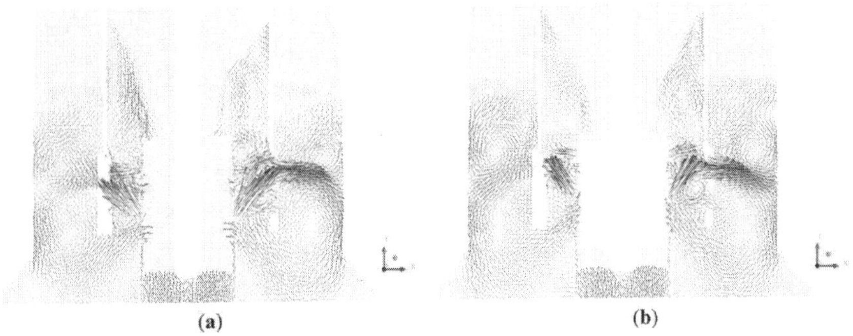

Figure 9. (**a**) Water superficial velocity vectors at maximum rate of air flow,d_b = 0.7 mm; (**b**) water superficial velocity vectors at minimum rate of air flow, d_b = 0.7 mm.

5. CFD-BASED FLOTATION MODEL

Collisions rate between particles and bubbles is a critical factor for minerals recovery by flotation machines. Collisions rate depends on the local turbulent dissipation rate in the pulp phase. Other factors such as attachment and detachment rate of particles-bubbles aggregates depend on turbulent dissipation rate as well as surface chemistry of particles and bubbles. Two-phase flow simulations of flotation machines provide spatial distribution of air void fraction and turbulent dissipation rate. A CFD-based flotation model has been developed in this work to predict pulp rate constant of Wemco 0.8 m^3 pilot cell. The model provides local values of collisions rate between particles and bubbles as well as probabilities of attachment, collisions and stabilization. Abrahamson collision [8] model is commonly used in estimation for collisions rate in flotation modeling. This paper provides an alternative more accurate model for collisions rate developed by Zaichik *et al.* [9]. In this paper, we present flotation model as a first rate equation. The rate constant in this equation depends on the local collisions frequency, probability of collision, probability of attachment, and probability of stabilization.

Yoon and Lutrell [10], Pyke *et al.* [11], Bloom and Heindel [12] among others developed theoretical flotation models to predict pulp recovery rate. In these models, flotation process is represented mathematically by a first order kinetic ODE Ordinary Differential Equation). These theoretical models depend on the average air void fraction and dissipation rate for a given flotation cell. However, distributions of turbulent dissipation rates and air volume fraction are not uniform and depend on the design of the machine. Distribution of turbulent dissipation rate is the key factor in computing recovery rate. Koh and Schwarz [13] developed a 3D CFD flotation model for a forced air Rushton's impeller flotation tank. Collisions frequency and flotation probabilities have been computed locally. CFD simulations of flotation machines provide the spatial distribution of the rate of turbulent kinetic energy dissipation and air void fraction throughout the cell. Flotation model is based on estimating local recovery rate. Local recovery rate is the multiplication of collision kernel and probabilities of collision, attachment and stabilization. These parameters depend mainly on dissipation rate throughout the machine in addition to particles and bubbles sizes, surface tension and contact angle. The model has been applied to a self-aerated flotation machine (Wemco 0.8) to predict pulp recovery rate. In this paper, a range of particle sizes ($10\ \mu m \leq d_p \leq 500\ \mu m$) have been used with three different bubble sizes $d_b = 0.5$ mm, $d_b = 0.7$ mm, $d_b = 1.0$ mm to investigate the effects of particles and bubbles sizes on pulp rate constant. Effects of feed particles size distribution has been studied also in this work.

5.1. Flotation Model: First-Order Rate Equation

The objective of a flotation model is to predict the recovery rate (rate of mass flow of useful minerals collected from a flotation cell).

Following Koh and Schwarz [13], we model the flotation kinetics as a first-order rate process, which is given by the fundamental equation:

$$\frac{dN_{\text{p1}}}{dt} = -K_1 N_{\text{p1}} N_{\text{b}} + K_2 N_{\text{a}} \tag{5}$$

where Np1 is the particle number concentration (number of particles per unit volume) of free particles (not attached to bubbles), Nb is the number concentration of bubbles available for attachment (not fully loaded bubbles), Na is the number of particle-bubble aggregates (bubbles that cannot accept more particles but can lose particles), K1 is the average particle-bubble attachment rate constant, and K2 is the average particle-bubble detachment rate constant. We note the difference in the dimensions (units) of K1 (m^3/s) and K2 (1/s). The total particle number concentration NPT is:

$$N_{PT} = N_{\text{p1}} + N_{\text{p2}} \tag{6}$$

where Np2 is the number concentration of particles attached to bubbles. All number concentrations are functions of time and location in a flotation cell. A bubble is either fully loaded (cannot accept more particles) or clean (no particles attached). The number of particle-bubble aggregates is proportional to the total number of bubbles,

$$N_{\text{a}} = b\, N_{bT} \tag{7}$$

where b is an average loading parameter, which varies with time and position. The number of clean bubbles (those available for attachment) is

$$N_{\text{b}} = (1 - b) N_{bT} \tag{8}$$

The rate equation can now be written as:

$$\frac{dN_{p1}}{dt} = -k_1(1-b)N_{p1}N_{bT} + k_2 bN_{bT} \tag{9}$$

The number of particles (S) that can be attached to a bubble is given by the ratio of surface area of the bubble to the projected area of the particle.

$$S = 4\left(\frac{d_b}{d_p}\right)^2 \tag{10}$$

Such an estimation is not realistic. In Koh and Schwarz [13] work, only half of that number is assumed as a first approximation.

$$S_{max} = 0.5\,S = 2\left(\frac{d_b}{d_p}\right)^2 = \frac{N_{p2}}{\beta\, N_{bT}} \tag{11}$$

Rearranging, we obtain:

$$b = \frac{N_{p2}}{S_{max}\, N_{bT}} = \frac{N_{p2}}{2\, N_{bT}}\left(\frac{d_b}{d_p}\right)^2 \tag{12}$$

5.2. Attachment Model

Collisions frequency (Zpb) of two groups of dispersed species (e.g., particles and bubbles) is the number of collisions per unit volume per unit time ($m^{-3}\cdot s^{-1}$). It is proportional to the product of the number densities N_b and N_p of bubbles and particles (Zpb=β Np Nb), where β is called the collisions kernel (m3/s). The attachment rate constant (k1, m3/s) is defined as the product of particle-bubble collisions kernel times probabilities (or efficiencies). This is because not every collision event leads to successful attachment of a particle to a bubble. Following Koh and Schwarz [13], the attachment rate constant is written as:

$$k_1 = \beta \, P_c \, P_a \, P_s \tag{13}$$

where P_c, P_a, and P_s are probabilities of particle-bubble collision, adhesion, and stabilization. Multiplication by $P_c$is questionable since it should be considered as a part of the collisions kernel. Apparently including P_c is a remnant of the original model that was developed under quiescent conditions. This collisions probability $P_c$should be eliminated because collisions kernel is modeled under turbulent conditions, and the assumption of laminar flow is irrelevant. When turbulence is the mechanism of collisions of particles and bubbles, "quiescent" conditions should be defined as the turbulent fluctuations u' and turbulent dissipation rate that maximizes the attachment rate and minimizes detachment.

5.3. Collision Kernel

The collision kernel β is perhaps the most important ingredient of the attachment rate constant k1. For turbulent flows, two classical models are in use. The first is Saffman and Turner model [14], which is applicable in the limit of zero Stokes particle number. The second is Abrahamson model [8] which is applicable in the limit of infinite particle Stokes number. The form of the later model is usually written (Schubert and Bischofberger 1979) [15] as:

$$\beta = 5\left(\frac{d_p + d_b}{2}\right)^2 \sqrt{U_p^2 + U_b^2} \tag{14}$$

where Up and Ub are the turbulent (rms) fluctuating velocities of the particles and bubbles relative to the carrier liquid, respectively. Leipe and Mockel's [16] formula (not the original formula by Abrahamson) is used for these velocities,

$$U_i = \frac{0.4\,\epsilon^{4/9} d^{7/9}}{\nu^{1/3}}\left(\frac{|\rho_i - \rho_f|}{\rho_f}\right)^{2/3} \tag{15}$$

where ϵ is dissipation rate of the turbulent kinetic energy per unit mass (w/kg), ν and ρf are the kinematic viscosity and density of the liquid, respectively. ρi and di are the density and diameter of the colliding particles,i = p for particles and i = b for bubbles. The mass density for a bubble is assumed to be equal $0.5\rho f$, as if the bubble mass is the virtual mass of a spherical bubble. This equation implies that high dissipation results in higher collisions rates between bubbles and particles, and in this regard high dissipation has favorable effects on flotation. Schubert [15] model have been developed for very high inertia particles. In minerals, flotation particles have a wide spectrum of inertia. Therefore, application of Schubert collisions model results in over prediction of collision kernel as discussed by Fayed [17].

Zaichik *et al.* [9] developed another statistical model for the collisions kernel for particles and bubbles. The model applies to arbitrary values of density ratio and particle sizes. This model has been validated by Fayed and Ragab [18] to study its limitation and it provides more accurate prediction for collisions kernel than Schubert model [15]. The model is not presented here because of the space and interested reader is referred to Zaichik *et al.*[9] and Fayed [17] where the model has been presented and discussed extensively. In this paper, comparison between Schubert model [15] and Zaichik *et al.* model [9] is presented to show the over-estimation of collision kernel by the former model. Hence, Zaichik *et al.* model [9] has been used herein for more accurate prediction of collision kernel.

5.4. Flotation Probabilities

Process of minerals flotation involves two main events which are collisions of particles with bubbles and attachment of the colliding particles with bubbles. The first event happens under turbulent flow conditions where particles and bubbles sizes and turbulent velocity *rms* fluctuations are the main variables that control this event. Upon collision, useful minerals particles have to attach to the bubble and then transported to the froth phase by buoyancy of bubbles. The attachment process is due to the adhesive forces between particles and bubbles and then stabilization of the particle-bubble aggregates. An accurate model for collisions frequency of particles and bubbles has been used in this work [9] to estimate kernel collisions. The second step is to model for the probability of particles' attachment to a bubble. In the literature, considerable efforts have been made to model for the probabilities of collisions, adhesion and stabilization. The concept of probability of collisions has been developed under laminar flow conditions which do not exist in minerals flotation. In this section, we summarize and present models for probabilities of collisions, adhesion and stabilization developed by other authors. These models have been used in our CFD-flotation model to predict pulp recovery rate of a flotation cell. Hopefully, some validation work will be completed on a fundamental level in future to study the limitations of these models and their accuracy.

The probability (efficiency) of collision P_c is given by a formula due to Yoon and Luttrell [10],

$$P_c = \left(\frac{3}{2} + \frac{4}{15} {Re_b}^{0.72}\right) \frac{{d_p}^2}{{d_b}^2} \tag{16}$$

where the bubble Reynolds number is defined by Reb = db Ubυ. In the actual calculations Pc is limited to a maximum of 1.

The expression for probability of adhesion P_a is also derived by Yoon and Luttrell [10],

$$P_a = \sin^2\left(2\tan^{-1}\exp\left[\frac{-\left(45 + 8\,{Re_b}^{0.72}\right) U_b\, t_{\mathrm{ind}}}{15\, d_b \left(\frac{d_b}{d_p} + 1\right)}\right]\right) \tag{17}$$

where tind is the induction time, which is determined by an empirical formula due to Dai *et al.* [19]:

$$t_{\text{ind}} = \frac{75}{\theta}\, \mathrm{d}_p^{0.6} \tag{18}$$

where tind is measured in seconds, θ is particle-bubble contact angle in degrees, and dp is the particle diameter in meters.

The formula proposed by Schulze [20] and modified by Bloom and Heindel [12] for the probability of stabilization Ps is used:

$$P_s = 1 - \exp\left[A_s \left(1 - \frac{1}{\min(1, \mathrm{Bo}^*)}\right)\right] \tag{19}$$

where the modified Bond number is defined by:

$$\mathrm{Bo}^* = \frac{\mathrm{d}_p^2 \left[\Delta\rho g + 1.9\, \rho_p\, \epsilon^{\frac{2}{3}} \left(\frac{\mathrm{d}_p}{2} + \frac{\mathrm{d}_b}{2}\right)^{\frac{-1}{3}}\right] + 1.5\, \mathrm{d}_p \left(\frac{4\sigma}{\mathrm{d}_b} - \mathrm{d}_b \rho g\right) \sin^2\left(\pi - \frac{\theta}{2}\right)}{\left|6\sigma \sin\left(\pi - \frac{\theta}{2}\right) \sin\left(\pi + \frac{\theta}{2}\right)\right|} \tag{20}$$

where A_s = 0.5 is an empirical constant suggested by Bloom and Heindel [12], where σ is the surface tension (N/m), $\Delta\rho = \rho_p - \rho_f$, and g is the gravitational acceleration.

5.5. Detachment Model

The particle-bubble detachment rate constant (k2) is defined by:

$$k_2 = Z_2\, P_d = Z_2(1 - P_s) \tag{21}$$

The probability of detachment Pd is assumed to be equal to (1–Ps). Koh and Schwarz [13] justified the inclusion of Ps in both k1 and k2 because the processes involve different turbulent eddies acting independently of each other. The detachment frequency Z2 is given by Bloom and Heindel [12] as:

$$Z_2 = \frac{\sqrt{C_1}\ \varepsilon^{\frac{1}{3}}}{\left(d_p + d_b\right)^{\frac{2}{3}}} \tag{22}$$

where C1=2.0 is an the empirical constant. We note in this equation the unfavorable effects of dissipation that high dissipation results in destabilization of particle-bubble aggregates.

5.6. Particle Size Distribution

Particles in the feed slurry to a flotation machine have a particle size distribution. Several models in the literature have been developed to account for different particles properties in the feed slurry. These models predict the global rate constant of a flotation cell and classified into three groups—discrete rate constant distribution, continuous rate constant distribution and mean rate constant. Several discrete rate constant models have been developed but they differ in the number of fractions (Morris [21], Kelsall [22], Cutting [23], Jowett [24], Imaizumi [25]). Two fraction discrete models developed by (Kelsall [22]) are named as fast float and slow float fractions. Continuous rate constant models assume that rate constants are distributed as a continuous distribution represented by a Gamma-distribution function (Harris [26], Woodburn [27], Loveday [28], Kappur [29]). The mean rate constant model has been proposed by Chen (Chen Z.M. [30,31]). However, prediction of rate constant in all flotation models is strongly affected by operating conditions as well as particles and bubbles properties on a local basis. Flotation modeling using CFD enables more accurate prediction of rate constant and helps to assess the performance of a flotation cell.

The attachment rate k1 as given above is mainly a function of bubble size, particle size, air/water surface tension and contact angle. An effective pulp recovery rate constant proposed here is defined by:

$$k_1^* = k_1 \frac{\alpha}{v_b}(1 - \alpha) \tag{23}$$

We recognize the factor αvb as the local number concentration of bubbles, and the factor (1−α) is included so that in a region of high air volume fraction (α≈1) the recovery rate should drop to zero. The local number concentration of particles Np depends on slurry loading, and we expect it to be proportional to the local water volume fraction, (1−α). We believe that including the factor (1−α) gives a better figure of merit that can be used to evaluate different designs in the absence of Np distribution. The machine average rate constant is obtained by integration over the machine volume,

$$K_1^* = \frac{R_f}{V} \int_V k_1^* (1 - b) dV \tag{24}$$

where Rf is a froth recovery factor.

Minerals particles in the feed slurry have a size distribution and this distribution has great impact on recovery rate. Let P(dp) denote the pdf of particle size distribution. Particle size distribution depends on the grinding process and metallurgical properties of the minerals particles. The current CFD-based flotation kinetic model can be used to determine pulp recovery rate for both bubble size and particle size distributions. We assume a Gamma-distribution for the particles number density function.

$$P(d_p) = \frac{(d_p - \mu_p)^{\gamma - 1}}{B^\gamma \, \Gamma(\gamma)} \exp\left(-\frac{d_p - \mu_p}{B}\right) \tag{25}$$

where μp is the minimum possible particle diameter and Γ is the Gamma function. In this paper, we used μp=0and B = 30. To obtain different particles number distributions, we assign different values for γ starting from 1.0–15, incremented by one. This also changes the mean diameter. The pseudo rate constant for particle size distribution $\overline{K_1^*}$ is computed:

$$\overline{K_1^*} = \int_{d_{p,min}}^{d_{p,max}} P(\lambda) K_1^*(\lambda)\, d\lambda \tag{26}$$

and dp, min, dp, max are the minimum and maximum particle diameters present in the feed slurry.

5.7. Flotation Results

In this section, we demonstrate the viability of CFD-based flotation model as a tool to evaluate the performance of flotation machines and provide detailed hydrodynamic and kinetics data that can help improve the design of such machines. Ragab and Fayed [32] developed a CFD-based flotation model, and used it to determine the effects of particle size on the rate constant. We use particle specific gravity ρ=4.1, contact angle θ=40°, and surface tension σ=0.06 N/m as input parameters to the flotation model. Two-phase hydrodynamic simulations provided spatial distributions of dissipation rate, ϵ, and air volume fraction, α (also called void fraction). The number concentration of bubbles is NbT=α/vb, where vb is the bubble volume and spherical bubbles are assumed. Particles-bubbles collisions rate has been estimated using Zaichik *et al.*'s model [9]. Zaichik*et al.*'s model and probabilities models rely on velocity fluctuations, bubble diameter, particle diameter, particle density, air void fraction, contact angle and air-water surface tension coefficient. Velocity fluctuation, u' has been estimated from the local eddy viscosity and dissipation rate. Local air void fraction is obtained from the CFD results of the two-phase flow. Other parameters are user input. CFD-based flotation kinetics model is a post processing program to the two-phase simulations runs of flotation cells such as Wemco 0.8 m^3. Effects of particles size distribution on pulp recovery rate are presented here.

As described above, accurate estimation of particle-bubble collision rate is essential for accurate modeling of rate constant. Fayed and

Ragab [18] validated Zaichik *et al.* model [9] and recommended it for use in flotation modeling. A comparison between collisions kernel by Schubert model [15] and Zaichik *et al.* model [9] are shown in Figure 10a,b. Schubert model overestimates the collisions kernel by an order of magnitude when compared to that of Zaichik *et al.* model. Based on these results, we recommend Zaichik *et al.* model in flotation kinetics. Figure 11a–c shows the spatial distribution of probabilities of collision, adhesion and stabilization in a vertical plane that passes through the machine axis. Figure 11a,b shows the favorable effects of turbulent dissipation rate where higher collisions rate and higher probability of collisions in the region between rotor and disperser and within the jet region. However, high local dissipation rates have adverse effects on the probabilities of adhesion and stabilization where low values are observed in the regions between rotor and disperser and within the jet. As described above, flotation process is mainly due to two events—collisions of particles and bubbles and particles attachment to a bubble. That is to say, we can achieve high dissipation rate regions that cause a high collisions rate but lower rates of attachment are obtained which results in a lower rate constant. Therefore, the presented CFD-based flotation model in this paper is a useful tool to optimize for rate constant in flotation machines.

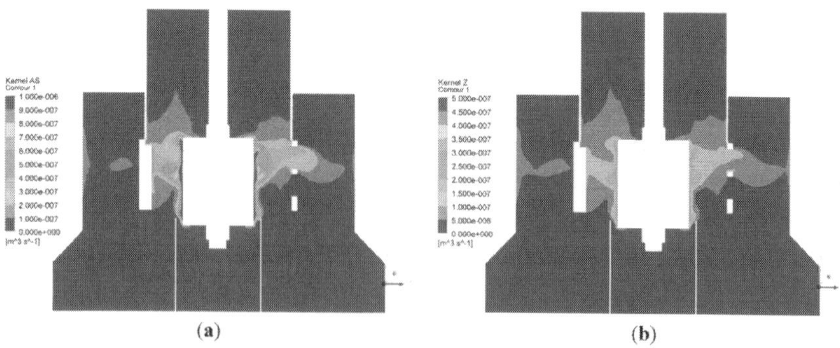

Figure 10. (**a**) Abrahmson-Schubert [6] collisions kernel for $d_p = 100$ micron; (**b**) Zaichik *et al.* [9] collisions kernel for $d_p = 100$ micron.

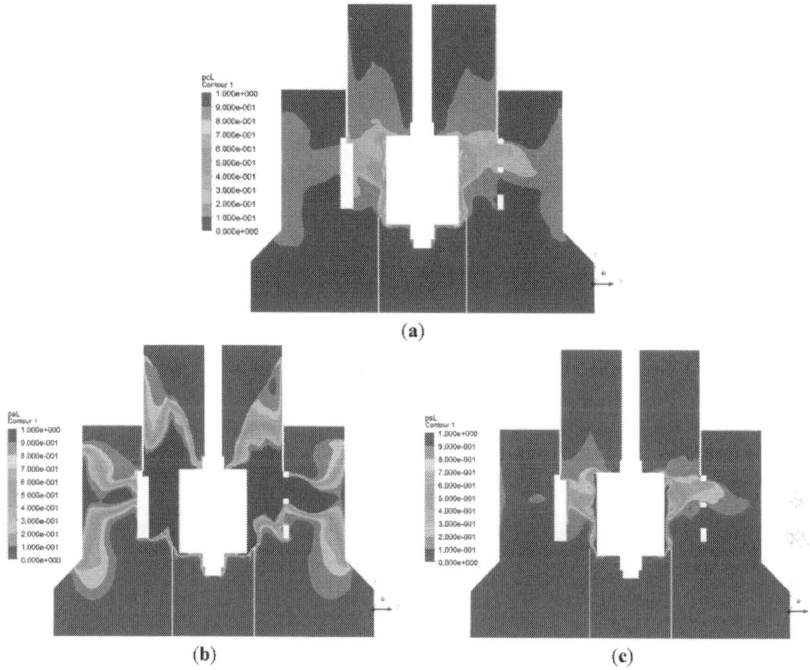

Figure 11. (**a**) Probability of collisions for d_p = 100 micron; (**b**) Probability of adhesion for d_p = 100 micron; (**c**) Probability of stabilization for d_p = 100 micron.

The product of collision kernel and probabilities of collision, adhesion and stabilization gives the local attachment rate k1. An effective pulp recovery rate constant (in the absence of detachment) is defined here byk1α (1−α)/vb, and corresponding contours are depicted in Figure 12a–d. The factor (1−α) is included in this definition so that in a region of 100% air (α=1.0) the local recovery rate should be zero. The spatial distribution of α(1−α), which is zero if α=0.0 or α=1.0 and maximum if α=0.5 (50% void fraction), implies that the local recovery rate is maximized when the air is well dispersed throughout the machine. The actual recovery rate will depend on the balance between the favorable effects of dissipation rate ϵ in increasing the collision frequency against its adverse effects on reducing the attachment rate and increasing detachment rate.

Therefore, knowing the spatial distributions of both α and ϵ throughout the machine is essential in understanding the effectiveness of different components (rotor, stator or disperser, jets) on the flotation efficiency. Only through CFD of two-phase simulations (or elaborate experimental measurements) one can determine those spatial distributions. This is a clear advantage of CFD-based flotation models in comparison with models that treat the entire cell as one unit which assumes a single value for ϵ determined by the consumed power and single value for α that is equal to the gas holdup.

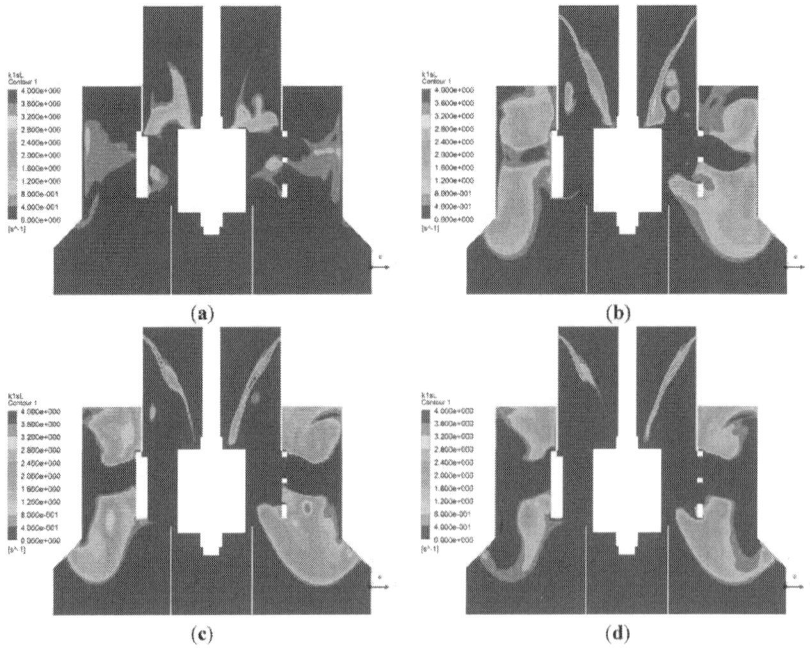

Figure 12. (**a**) Pulp recovery rate k∗1 for 40 μm particle diameter; (**b**) Pulp recovery rate k∗1 for 100 μm particle diameter; (**c**) Pulp recovery rate k∗1 for 200 μm particle diameter; (**d**) Pulp recovery rate k∗1 for 300 μm particle diameter.

Particle diameter has significant effects on distribution of the pulp rate constant k∗1 throughout the Wemco machine. Contours of k∗1 in a vertical mid plane are shown in Figure 12a–d, for particle diameters of 40, 100, 200, and 300 μm. Fine particles dp< 50 μm are efficiently recovered in the high dissipation region between the rotor and disperser. For a medium particle diameter of 100 μm≤dp≤200 μm, recovery happens in the jets out of the disperser. For coarse particles dp≥ 300 μm pulp recovery is more efficient in the moderate dissipation regions outside the disperser below, above and between the disperser jets. In the present work, the pseudo rate constant is computed as a function of particle diameter for two values of contact angle (θ=40° and θ=50°). Values for rate constant are obtained for different uniform bubble size and different uniform particle size. Average flotation rate constant k∗1 can be defined according to the Equation (24) where Rf is a froth recovery factor. We assumed Rf=1.0 and β=0.0 for a fully unloaded bubble and studied the effects of bubble diameter, particle diameter, and contact angle on k∗1. The results are shown in Figure 13a–c. As expected, rate constant increases with the increase of contact angle. Also, higher rate constant is observed for smaller bubble size (db=0.5 mm) because air holdup increases for smaller bubble diameter, and hence higher bubble concentration number, as well as a lower rate constant for larger bubble diameter due to the decreases in the air holdup (*i.e.*, local bubbles number concentration). The maximum recovery rate shifts to higher particle diameter with the increase in bubble diameter. The recovery rate shown in these figures is very high relative to experimental data reported in literature because of the assumed values of Rf and β. Maximum rate constant for bubble sizes of 0.5, 0.7 and 1.0 mm occur at particle size of 125, 150 and 175 μm, respectively. This reveals that fact that smaller bubbles sizes (d_b < 0.5 mm) are needed to float very tiny particles (d_p < 100 μm).

Next, we study the effects of particles size distribution on the mean flotation rate constant. As shown in Figure 13a–c, rate constant is dramatically affected by particle sizes and the feed slurry to a flotation cell contains a wide spectrum of particles sizes. Therefore, modeling for rate constant with particle size distribution is more realistic and useful. Samples of assumed particles' number distribution are shown in Figure 14 for B = 30 and three different values for γ (γ = 2, 3 and 5). Changing values for γ changes the size distribution and mean diameter of the particles. In real samples, this distribution depends on the grinding process as well as metallurgical properties of the ore. The procedure herein is to apply the present CFD flotation model for a single particle size and uniform bubble diameter in the postprocessor of the used CFD package (CFX). Using Equation (23), we obtain the local rate constant for a specific particle size. We used particles sizes ranging from 10–500 μm to be within the practical particle sizes. We used Equation (24) to obtain the average rate constant over the pulp volume and from Equation (26) we calculate the average pulp rate constant for a specific distribution. This process yields a single value for the rate constant that we attribute to the mean value of the proposed particle size. Different particle number distributions pdfs are used (each has different mean diameter) to plot rate constant *versus*particles mean diameter. This deterministic method in the calculation of rate constant for certain particle size distribution is more robust than the other probabilistic models [26,27,28,29]. This is because the effects of all variables such as surface tension, contact angle, bubbles size and particles size are explicitly included in the model.

The effect of particles size distribution on the rate constant has been studied also by an assumed γ-particle number distribution. Effects of different size distribution on mean rate constant are depicted in Figure 15a–c for different bubbles sizes. Figure 15a shows that the maximum rate constant is around the mean particle diameter of 140 μm for a contact angle of 50°, and the particle diameter of 120 μm for a contact angle of 40° where, increasing contact angle increases the possibilities of floating larger particles. A considerable mean rate constant for very small mean diameter (d_{mean} = 5 μm) is observed for bubble size of 0.5 mm while very low rate constant exists for the larger bubble size (d_{mean} = 0.7 and 1.0 mm) at the same particle mean diameter. Maximum mean rate constant for bubble size (d_{mean} = 0.7 mm) is also observed to be around particle mean diameter 140 and 180 μm for bubble size (d_{mean} = 1.0 mm) as shown in Figure 15b,c.

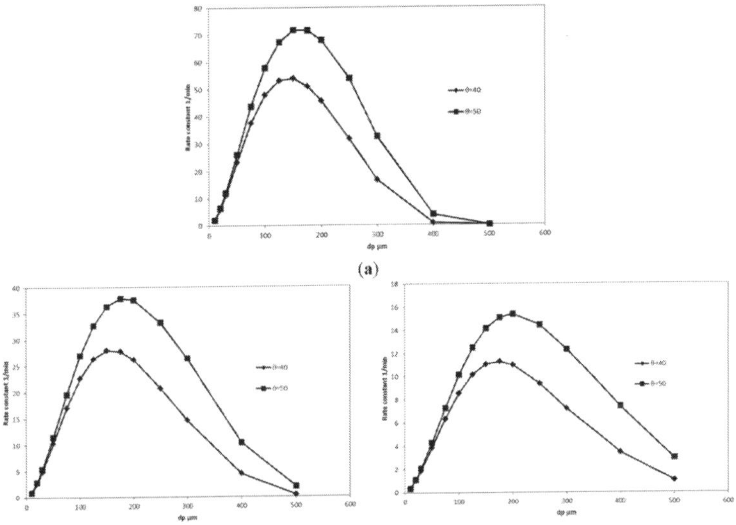

Figure 13. (**a**) Average rate constant for 0.5 mm bubble diameter and for single particle sizes; (**b**) average rate constant for 0.7 mm bubble diameter and for single particle sizes; (**c**) average rate constant for 1.0 mm bubble diameter and for single particle sizes.

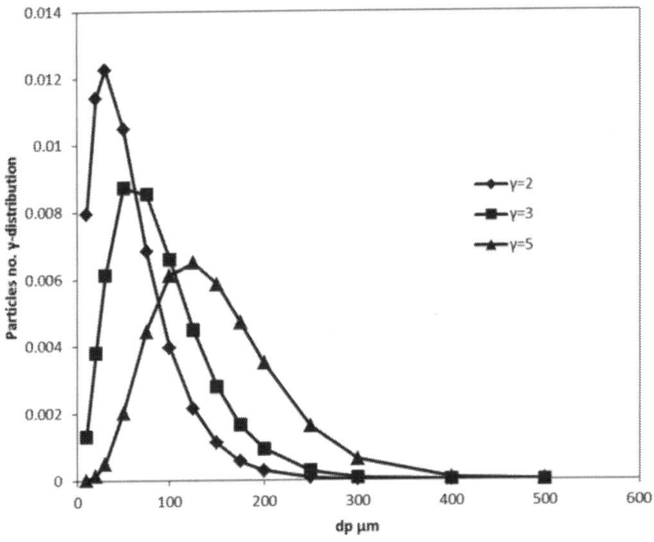

Figure 14. Assumed particles number distribution, $B = 30$.

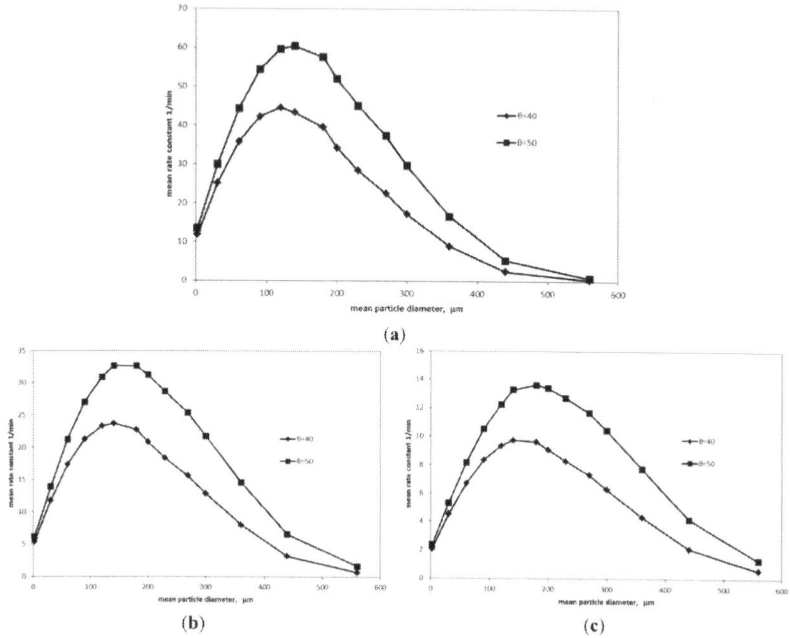

Figure 15. (**a**) Average rate constant for 0.5 mm bubble diameter and for particle no. γ-distribution; (**b**) average rate constant for 0.7 mm bubble diameter and for particle no. γ-distribution; (**c**) average rate constant for 1.0 mm bubble and diameter for particle no. γ-distribution.

6. CONCLUSIONS

Two phase simulations of self-aerated flotation Wemco 0.8 m^3 pilot cells have been conducted to study the transient flow structures and predict air flow rate and power consumption as a function of time. The newly devised overflow tank on top of the pulp volume and constant static pressure at the top boundary of this tank and the standpipe opening allowed us to predict the airflow rate and the transient position of the pulp–air interface until it reached a steady position. Also, investigating dynamical behavior of two-phase flow in the standpipe provides us with a detailed picture of the mixing of air and water and pumping this mixture into the tank. The two-phase simulations provide us with essential parameters for flotation modeling such as local turbulent dissipation rate, velocity *rms* fluctuations and air void fraction throughout the whole machine. These parameters have been used to develop a CFD-based flotation model. This model predicts pulp rate constant for an arbitrary particle size and bubble size. It uses a first-order rate equation, where processes of collision, attachment and detachments are described by well-known theoretical and empirical formulae. The model uses local values of the rate of turbulent energy dissipation and air volume fraction. Not only the average pulp recovery rate can be estimated but also the regions of high/low recovery rate can be identified. The CFD-based flotation model presented here is also used to determine the dependence of recovery rate constant at any locality within the pulp on bubble diameter, particle diameter, particle specific gravity, contact angle, and surface tension. Furthermore, we have updated our CFD-based flotation kinetics model to predict pulp recovery rate in the presence of both particles size distributions. The calculations are repeated for many particle diameters selected from a range that covers the anticipated minimum and maximum diameters in the slurry. The particles number density pdf and the data generated for single particle size are used to compute the recovery rate for a specific mean particle diameter. Our computational model gives a figure of merit for the recovery rate of a flotation machine, and as such can be used to assess incremental design improvements and design of new machines. The model is a very useful design tool because it can be used to establish the effects of different components (rotor, stator, or disperser jets) on the pulp recovery.

ACKNOWLEDGMENT

This work has been supported by FLSmidth Minerals Inc., Salt Lake City, Utah.

AUTHOR CONTRIBUTIONS

Dr. Fayed and Dr. Ragab worked closely on the material presented in this paper during the time Dr. Fayed was working as Graduate Research Assistant in Department of Engineering Science and Mechanics, Virginia Tech, USA. Dr. Fayed constructed geometric models of Wemco machine, generated the mesh, and conducted computer simulations. Dr. Ragab contributed to the theoretical modeling and boundary conditions. Both authors contributed equally to data analysis and drawing conclusions.

REFERENCES

1. Koh, P.T.L.; Schwarz, P. CFD Model of a Self-Aerating Flotation Cell. *Int. J. Miner. Process.* **2007**, *85*, 16–24.

2. Tiitinen, J.; Koskinen, K.; Ronkainen, S. Numerical modeling of an Outokumpu flotation cell. In Proceedings of the Centenary of Flotation Symposium, Brisbane, Australia, 6–9 June 2005.

3. Kerdouss, F.; Bannari, A.; Proulx, P. CFD modeling of gas dispersion and bubble size in a double turbine stirred tank. *Chem. Eng. Sci.* **2006**, *61*, 3313–3322.

4. Prosperetti, A.; Tryggvason, G. *Computational Methods for Multiphase Flow*; Cambridge University Press: Cambridge, UK, 2007.

5. Clift, R.; Grace, J.R.; Weber, M.E. *Bubbles, Drops and Particles*; Dover Publications: Mineola, NY, USA, 2005.

6. Van den Akker, H.E.A. Toward a truly multiscale computational strategy for simulating turbulent two-phase flow processes. *Ind. Eng. Chem. Res.* **2010**, *49*, 10780–10797.

7. Nelson, M.; Traczyk, F.; Lelinski, D. Design of Mechanical Flotation Machines. In Proceedings of the 2011 SME Annual Meeting, Denver, Colorado, USA, 27 February–2 March 2011.

8. Abrahamson, J. Collision rate of small particles in a vigorously turbulent fluid. *Chem. Eng. Sci.* **1975**, *30*, 1371–1379.

9. Zaichik, L.I.; Simonin, O.; Alipchenkov, V.M. Turbulent collision rates of arbitrary density particles. *Int. J. Heat Mass Transf.* **2010**, *53*, 1613–1620.

10. Yoon, R.H.; Luttrell, G.H. The effect of bubble size on fine particle flotation. *Miner. Process. Extr. Metall. Rev.* **1989**, *5*, 101–122.

11. Pyke, B.; Fornasiero, D.; Ralston, J. Bubble particle heterocoagulation under turbulent conditions. *J. Colloid Interface Sci.* **2003**, *265*, 141–151.

12. Bloom, F.; Heindel, T.J. Modeling flotation separation in semi-batch process. *Chem. Eng. Sci.* **2003**, *58*, 403–422.

13. Koh, P.T.L.; Schwarz, M.P. CFD modeling of bubble-particle attachments in flotation cells. *Miner. Eng.* **2006**,*19*, 619–626.

14. Saffman, P.G.; Turner, T.S. On the collision of drops in turbulent clouds. *J. Fluid Mech.* **1956**, *1*, 16–30.

15. Schubert, H. On the turbulence-controlled microprocesses in otation machines. *Int. J. Miner. Process.* **1999**,*56*, 257–276.

16. Leipe, F.; Mockel, O.H. Untersuchungen zum stoffvereinigen in ussiger phase. *Chem. Technol.* **1976**, *30*, 205–209. (In German)

17. Fayed, H. Particles and Bubbles Collisions in Homogeneous Isotropic Turbulence and Applications to Minerals Flotation Machines. Ph.D. Thesis, Virginia Tech, Blacksburg, VA, USA, 6 December 2013.

18. Fayed, H.E.; Ragab, S.A. Direct Numerical Simulation of Particles-Bubbles Collisions Kernel in Homogeneous Isotropic Turbulence. *J. Comput. Multiph. Flows* **2013**, *5*, 168–188.

19. Dai, Z.; Fornasier, D.; Ralston, J. Particle-bubble attachment in mineral flotation. *J. Colloid Interface Sci.* **1999**,*217*, 70–76.

20. Schulze, H.J. Flotation as a hetrocoagulation process: Possibilities of calculation the probability of flotation. In *Coagulation and Flocculation*; Dobias, B., Ed.; Marcel Dekker: New York, NY, USA, 1993; pp. 321–363.

21. Morris, T.M. Discussion of flotation rates and flotation efficiency. *Min. Eng.* **1952**, *4*, 794–798.

22. Kelsall, D.F.; Stewart, P.S. A critical review of applications of models of grinding and flotation. In Proceedings of the Symposium on Automatic Control Systems in Mineral Processing Plant, Brisbane, Australia, 17–20 May 1971; pp. 213–232.

23. Cutting, G.W.; Devenish, M.A. Steady-state model of froth flotation structures. In Proceedings of the AIME Annual Meeting, New York, NY, USA, 20 February 1975.

24. Jowett, A. Resolution of flotation recovery curves by a difference plot method. *Trans. IMM* **1974**, *70*, 191–204.

25. Imaizumi, T.; Inoue, T. Kinetic considerations of froth flotation. In Proceedings of the 6th International Mineral Processing Congress, Cannes, France, 26 May–2 June 1963; pp. 581–593.

26. Harris, C.C.; Chakravarti, A. Semi-batch froth flotation kinetics; species distribution analysis. *Trans. AIME***1970**, *247*, 162–172.

27. Woodburn, E.T.; Loveday, B.K. Effect of variable residence time data on the performance of a flotation system. *J. South Afr. Inst. Min. Metall.* **1965**, *65*, 612–628.

28. Loveday, B.K. Analysis of froth flotation kinetics. *Trans. IMM* **1966**, *75*, 219–225.

29. Kapur, P.C.; Mehrotra, S.P. Estimation of the flotation rate distributions by numerical inversion of the Laplace transform. *Chern. Eng. Sci.* **1974**, *29*, 411–415.

30. Chen, Z.M.; Wu, D.C. A study of flotation kinetics. *Nonferr. Met.* **1978**, *10*, 28–33.

31. Chen, Z.M.; Mular, L. A study of flotation kinetics—A kinetic model for continuous flotation. *Nonferr. Met.***1982**, *3*, 38–43.

32. Ragab, S.; Fayed, H. CFD-Based Flotation Model for Prediction of Pulp Recovery Rate. In Proceedings of the SME Annual Meeting and Exhibit, Seattle, WA, USA, 19–22 February 2012.

3

EVALUATION OF CHARACTERISTICS OF PHASE CHANGE HEAT TRANSFER IN ULTRAFINE CRYOPROBE

Junnosuke Okajima, Atsuki Komiya, Shigenao Maruyama

Institute of Fluid Science, Tohoku University, Sendai, Japan

ABSTRACT

To reduce the invasiveness of cryosurgery, a miniaturized cryoprobe is necessary. The authors have developed an ultrafine cryoprobe for realizing low-invasive cryosurgery by local freezing. The objectives of this study are to estimate the heat transfer coefficient and investigate the characteristics of the phase change heat transfer in the ultrafine cryoprobe. This cryoprobe has a double-tube structure consisting of two stainless steel microtubes. The outer diameter of the cryoprobe was 550 μm. The alternative Freon HFC-23, which has a boiling point of −82°C at 0.1 MPa, was used as a refrigerant. To evaluate the characteristics of boiling flow in the cryoprobe, the heat transfer coefficient was estimated. The derived heat transfer coefficient was higher than that obtained from the conventional correlation. Additionally, a bubble expansion model was introduced to evaluate the heat transfer mode of the phase change flow in the ultrafine cryoprobe. This model can estimate the liquid film thickness during the expansion of a single bubble in a microchannel. The experimentally measured wall superheat was much lower than that obtained from the model. Therefore, this result also implied that the heat transfer mode in the ultrafine cryoprobe should be nucleate boiling.

KEYWORDS

Phase Change Heat Transfer, Biomedical Application, Cryosurgery, Microchannel

1. INTRODUCTION

Cryosurgery is a surgical treatment that uses the characteristics of frozen biological tissue to remove undesirable tissues using a cooling device called a cryoprobe. Cryosurgery is less invasive and offers the advantages of low bleeding and a short recovery period [1].

Some cryoprobes have succeeded commercially, e.g., Accuprobe from Cryomedical Science [2], CRYOcare from ENDOcare [3], ERBOKRYO from ERBE [4], CryoHit from Galil Medical [5], and SurgiFrost from CryoCath Technologies [6]. Conventional cryoprobes, which are classified into two types depending on the cooling method used, are 3 - 8 mm in diameter. Hence, it has been difficult to treat small lesions, such as pigments on the skin, wrinkles around the eyes, and early breast cancer, by conventional cryoprobes.

Several cryoprobes have been proposed that reduce the level of invasiveness. Takeda et al. [7] developed a Peltier cryoprobe. By changing the electric current supplied to a Peltier module, the surface temperature of the cooling section can be controlled precisely. Aihara et al. [8] developed a flexible, long, slender cryoprobe with vacuum insulation. Boiling heat transfer of an impinging jet using liquid nitrogen was used as the heat transfer mechanism. Additionally, Maruyama et al. [9] developed a flexible cryoprobe that uses the Peltier effect. This cryoprobe consists of a flexible plastic tube. Additionally, miniaturized cryoprobes of around 1 mm diameter have been studied by Benita and Condé [10] and Zhang et al. [11] ; however, the cooling power of those cryoprobes was insufficient because the heat transfer rate decreased owing to the large ratio of the surface area to the volume. Generally, the miniaturization of channel size causes increasing the heat transfer coefficient and decreasing the heat transfer rate comparing to the normal size. Therefore, a mechanism to achieve the both of high heat transfer coefficient and transporting large amount of heat is required and the solution is using the phase change heat transfer.

To overcome this problem, the authors have studied the phase change heat transfer in a co-axial small double tube [12] and developed an ultrafine cryoprobe with boiling heat transfer in a microchannel [13] [14]. The outer diameter of this cryoprobe is 550 μm. The ultrafine cryoprobe may realize new cryosurgery for small lesions with minimum invasiveness, or in a blood vessel using a catheter. In previous study, the theoretical limitation of freezing size by the ultrafine cryoprobe [14] and the refrigerant state inside the cryoprobe [13] were discussed. To realize small-scale cryosurgery using the ultrafine cryoprobe, its cooling performance needs to be clarified. Especially, the heat transfer coefficient is the important parameter to evaluate the cooling performance. The objectives of this study are to estimate the heat transfer coefficient and evaluate the characteristics of the phase change heat transfer in the ultrafine cryoprobe by using the estimated heat transfer coefficient.

2. EXPERIMENTAL SYSTEM

2.1. Ultrafine Cryoprobe

Figure 1 shows the concept of the ultrafine cryoprobe, which consists of inner and outer tubes. The inner tube has a 150 μm outer diameter (OD) and a 70 μm inner diameter (ID). The outer tube has a 550 μm OD and 300 μm ID. Both tubes are made of stainless steel. HFC-23, which is an alternative Freon with a normal boiling point of −82.1°C, was used as a refrigerant. HFC-23 was transported to the inner tube in the liquid state.

The inner tube acts as a capillary for depressurization. Therefore, a large pressure drop occurs in the inner tube, and the temperature of the refrigerant decreases. Furthermore, the refrigerant expands upon exiting the inner tube. Here, the refrigerant changes to a two-phase flow, and the outer tube is cooled. The advantages of this cooling system are as follows:

1) The refrigerant can be transported from the reservoir to the cooling section. Therefore, a complex insulation system such as that used in liquid nitrogen cryoprobes is not required.

2) The ultrafine cryoprobe is cooled by boiling heat transfer. Therefore, a higher heat transfer coefficient than that of Joule-Thomson cryoprobes can be expected.

3) Generally, the cryoprobe is required to generate lower than −20°C in order to necrotize an affected area. The size of necrosis area and required heat flux can be predicted by solving the freezing phenomena of biological tissue.

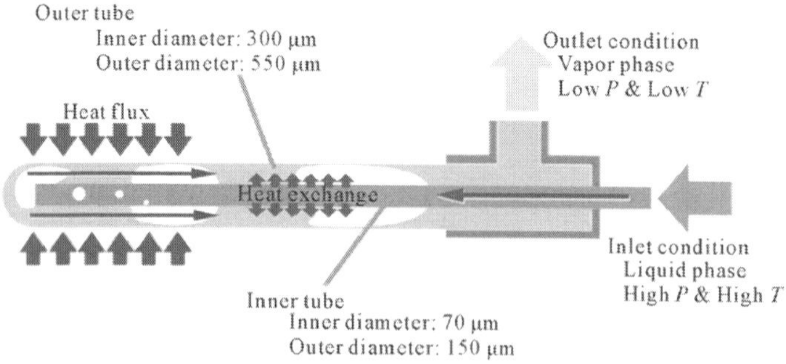

Figure 1. Concept of ultrafine cryoprobe.

2.2. Experimental System

Figure 2 shows the experimental apparatus, which consists of an HFC-23 cylinder, a precooler, several valves, and the cooling section. The measurements were made using T-type thermocouples, a pressure gauge, and a thermal mass flowmeter. The locations of each sensor are shown in Figure 2(b), which also shows the details of the ultrafine cryoprobe section. The refrigerant temperature and pressure at the inlet and outlet were measured by a T-type thermocouple 76 μm in diameter and the pressure transducer, respectively. Nine T-type thermocouples were installed on the ultrafine cryoprobe surface using a thermally conductive, electrically insulating adhesive. Moreover, a leading wire was soldered to the cryoprobe's surface to supply the electrical voltage for controlling the heat flux. The surface of the ultrafine cryoprobe was insulated by Styrofoam [Dow Chemical Company, thermal conductivity = 0.03 W/(m · K)].

Before the experiment, the air in the experimental apparatus was evacuated by a vacuum pump. During evacuation, the needle section was heated by applying a direct current to remove any remaining water vapor. The HFC-23 tank was kept at room temperature. The precooler was cooled by ice water. The HFC-23 vapor from the tank passed through the precooler, and HFC-23 condensed from the vapor phase to the liquid phase. Before the experiment began, Valve 2 was opened to bring the pressure inside the needle section to 0.1 MPa and release the pressure inside the cryoprobe. Then the experiment was started by opening Valve 1.

3. EXPERIMENTAL RESULTS

Figure 3 shows the time variation in the temperature and mass flow rate without heating by the electric current. The surface temperature of the ultrafine cryoprobe decreased immediately and then remained constant at around −50°C. In previous study [13], the vapor quality in the tip of cryoprobe was estimated as the slightly subcooled liquid. By assuming that the refrigerant in outer tube is saturated, the pressure in the tip of cryoprobe is estimated at around 0.5 MPa. Therefore, the pressure drop in the inner tube is 3.7 MPa. Additionally, the refrigerant temperature in the outlet decreased gradually. This result indicates that the refrigerant exhaust from the outer tube was capable of cooling.

Figure 4 shows the time variation in the surface temperature with a heat flux of 100 kW/m^2. The temperature increased by only 2 K after heating by the electric current. Figure 5 shows the time variation in the surface temperature with a heat flux of 170 kW/m^2. In this case, the temperature increased by 5 to 7 K after heating by the electric current. Additionally, the temperature response at TC9 was slightly fluctuated. This fluctuation indicated that the partial dryout should occur in the cryoprobe. These two figures proves that the phase change heat transfer of HFC-23 has a large cooling power.

Figure 6 shows the time variation in the surface temperature with a heat flux of 180 kW/m^2. The temperature increased by only 5 to 7 K after heating by the electric current. This trend was as same as the case of 170 kW/m^2. However, as shown in Figure 7, the unstable response of the time variation in the surface temperature was observed at TC8 when a heat flux of 180 kW/m^2 was supplied. These unstable phenomena occurred accidentally. As shown in Figure 7, the instability started on the downstream side. The temperature at TC9 was around −5°C. In this location, the refrigerant was completely superheated by the external heat flux. On the other hand, a constant temperature was maintained near the tip of the ultrafine cryoprobe. Therefore, the edge of liquid film should exist around the location of TC8 and the wetting and drying should occur periodically at this location.

This temperature fluctuation shown in Figure 7 should not affect to the freezing process in the cryosurgery. The temperature of frozen region should not follow the fluctuation because the heat capacity of frozen tissue is large. The problem in the real application is the refrigerant temperature and time-averaged heat transfer coefficient. Next, the heat transfer coefficient of the ultrafine cryoprobe was derived using the stable temperature data under heating.

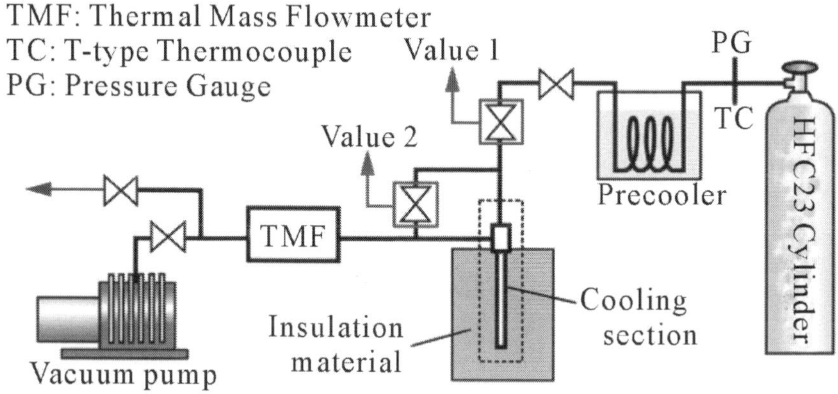

(a)

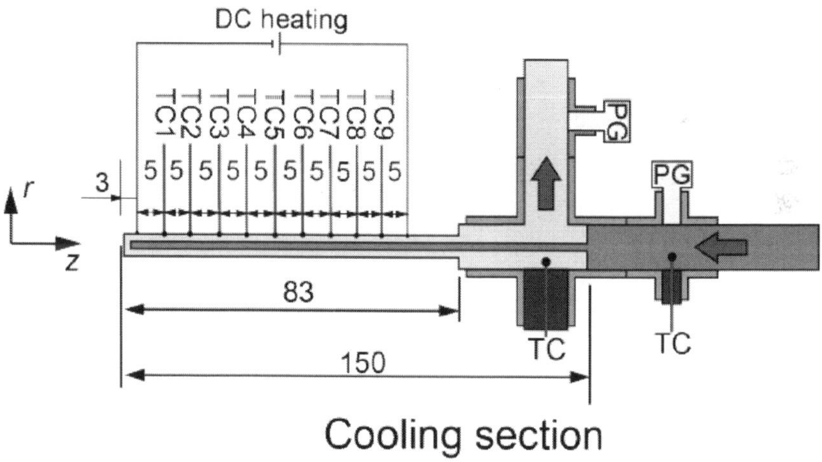

(b)

Figure 2. Schematic of experimental system, (a) Overview of system, (b) Detail of cooling section.

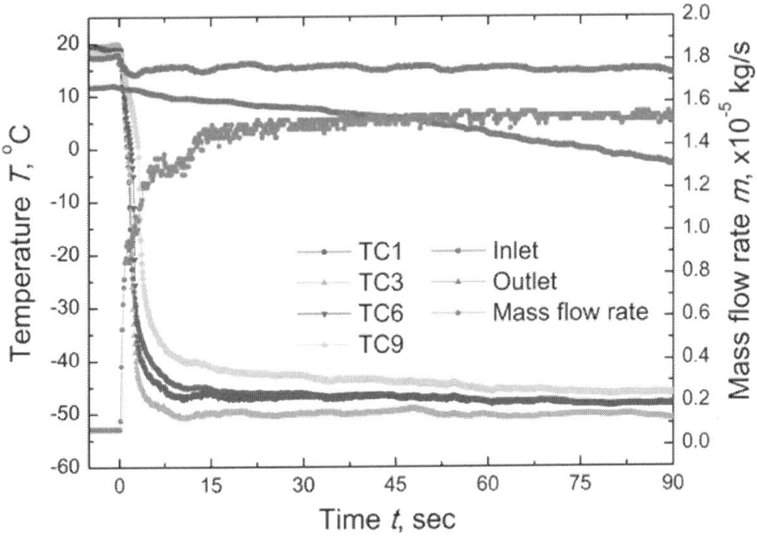

Figure 3. Time variation in temperature at TC1, TC3, TC6, TC9, inlet, and outlet and mass flow rate [14].

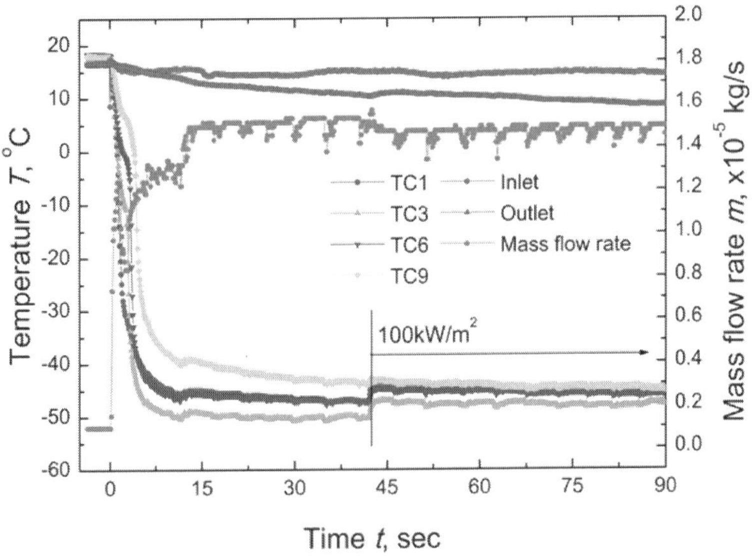

Figure 4. Time variation in temperature at TC1, TC3, TC6, TC9, inlet, and outlet and mass flow rate under heat flux of 100 kW/m^2 [14].

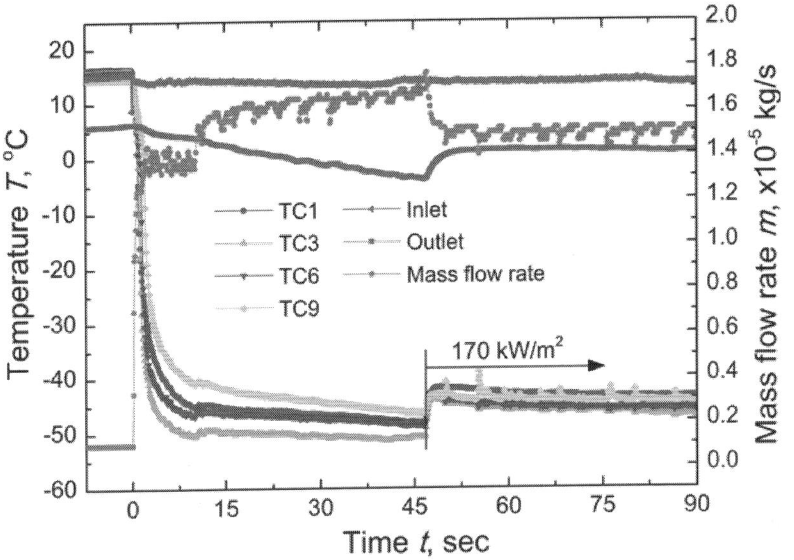

Figure 5. Time variation in temperature at TC1, TC3, TC6, TC9, inlet, and outlet and mass flow rate under heat flux of 170 kW/m[2].

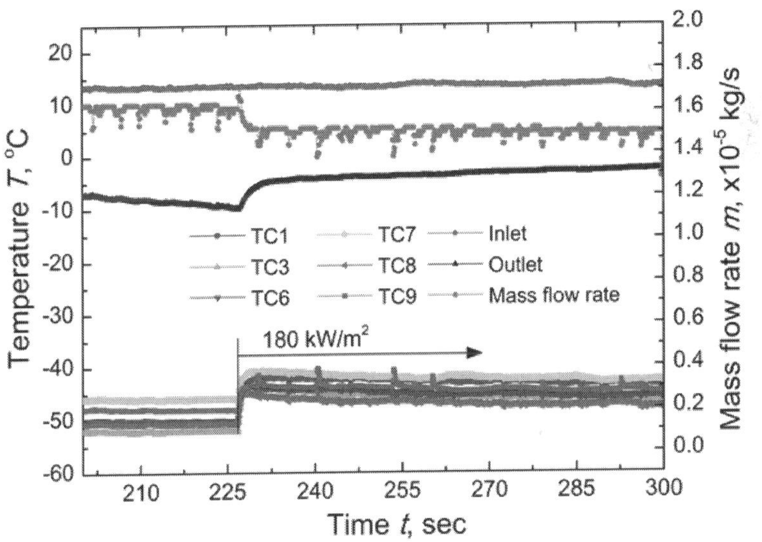

Figure 6. Time variation in temperature at TC1, TC3, TC6, TC7, TC8, TC9, inlet, and outlet, and in mass flow rate under heat flux of 180 kW/m[2].

4. HEAT TRANSFER COEFFICIENT

To consider the heat conduction in the channel wall, the one-dimensional axisymmetric heat conduction equation was solved. This equation with volumetric heat generation $\dot{q}_v$ in the steady state is expressed as

$$\frac{k_s}{r}\frac{\mathrm{d}}{\mathrm{d}r}\left(r\frac{\mathrm{d}T}{\mathrm{d}r}\right)=\dot{q}_v \quad (1)$$

The boundary conditions can be written as

$$\left.\frac{\mathrm{d}T}{\mathrm{d}r}\right|_{r=r_o}=0, \quad (2)$$

$$-k_s\left.\frac{\mathrm{d}T}{\mathrm{d}r}\right|_{r=r_o}=h\left[T_{sat}-T(r_i)\right]. \quad (3)$$

By solving Equation (1) analytically, the wall superheat and heat transfer coefficient can be expressed as

$$\Delta T_{sat}=T_{surf}-T_{sat}+\frac{q}{k_s}\left(\frac{r_i}{2}-\frac{r_i r_o^2}{r_o^2-r_i^2}\ln\frac{r_o}{r_i}\right), \quad (4)$$

$$h=\left[\frac{T_{surf}-T_{sat}}{q}+\frac{1}{k_s}\left(\frac{r_i}{2}-\frac{r_i r_o^2}{r_o^2-r_i^2}\ln\frac{r_o}{r_i}\right)\right]^{-1}, \quad (5)$$

where h: heat transfer coefficient [W/(m^2 · K)], k_s: thermal conductivity of stainless steel [W/(m · K)], q: heat flux [W/m^2], r_i: inner radius of outer tube[m] and r_o: outer radius of outer tube [m]. By using Equations (4) and (5), the wall superheat and heat transfer coefficient can be determined from the experimental data.

Figure 8 shows the wall superheat distribution on the cryoprobe surface derived using Equation (4). The wall superheat decreases slightly along the channel. This result indicates that heat conduction through the wall may affect the wall superheat. The heat flux of 150 kW/m^2 shows the highest wall superheat. Figure 9 shows the heat transfer coefficient distribution on the ultrafine cryoprobe surface derived using Equation (5). The heat transfer coefficient obviously increases along the channel, and the location changes dramatically at more than 30 mm. In this region, the axial heat conduction has a strong effect. Therefore, the representative wall superheat and heat transfer coefficient were determined by averaging the value sat locations of less than 30 mm.

Figure 10 shows the relationship between the heat transfer coefficient, wall superheat, and heat flux. Additionally, the estimated heat transfer coefficients are compared with Li's correlation [15], which is the correlation of the heat transfer coefficient of phase change flow in mini/microchannels, expressed as

$$Nu = 334 Bl^{0.3}\left(Bo \cdot Re_L^{0.36}\right)^{0.4}, \tag{6}$$

where k_L: thermal conductivity of the liquid phase [W/(m・K)], Nu: Nusselt number, Bl: boiling number, Bo: Bond number and Re_L: Reynolds number of the liquid phase, which are expressed as

$$Nu = \frac{hD}{k_L},\ Bl = \frac{q}{Gi_{LV}},$$
$$Bo = \frac{g\left(\rho_L - \rho_V\right)D^2}{\sigma},\ Re_L = \frac{G\left(1 - x_{eq}\right)D}{\mu_L}, \tag{7}$$

where G: mass flux [kg/(m^2・s)], i_{LV}: latent heat [J/kg], g: gravitational acceleration [m/s^2], ρ_L: density of the liquid phase [kg/m^3], ρ_V: density of the vapor phase [kg/m^3], σ surface tension [N/m], x_{eq}: vapor quality [-] and μ_L: viscosity [Pa・s]. As shown in

Figure 10, the heat transfer coefficient reaches a maximum value of 39 kW/(m^2 · K) at a heat flux of 170 kW/m^2. Furthermore, the heat transfer coefficient increases dramatically when the heat flux exceeds 150 kW/m^2. Compared with the value from Li's correlation, the experimental value was close to the heat transfer coefficient for low vapor quality under a low heat flux. However, in the high heat flux region, the rate of increase of the experimental results was larger than that from Li's correlation. This strong dependence on the heat flux is a general tendency of nucleate boiling. Therefore, Figure 10 implies that the heat transfer mode in the ultrafine cryoprobe is nucleate boiling.

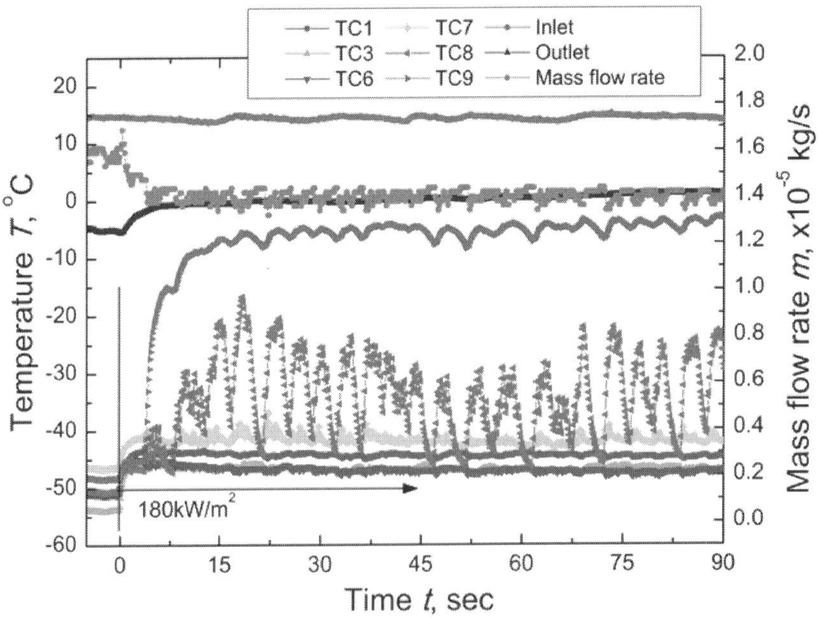

Figure 7. Unstable response in temperature at TC1, TC3, TC6, TC7, TC8, TC9, inlet, and outlet, and in mass flow rate under heat flux of 180 kW/m^2.

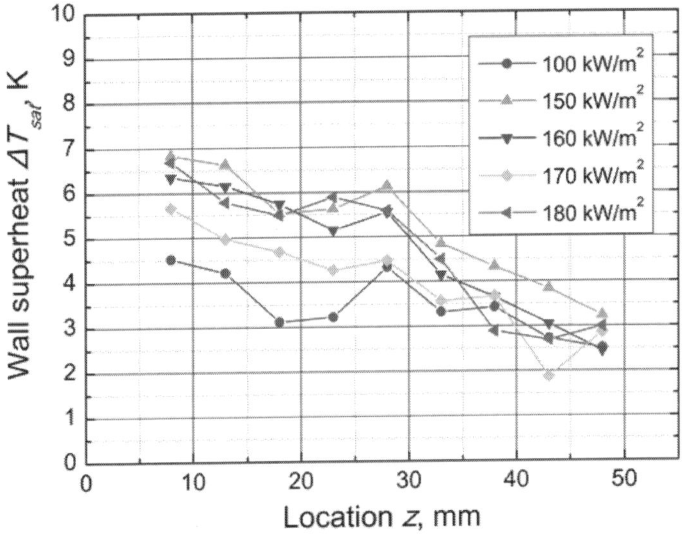

Figure 8. Wall superheat distribution on cryoprobe surface (Location z = 0 represents tip of cryoprobe).

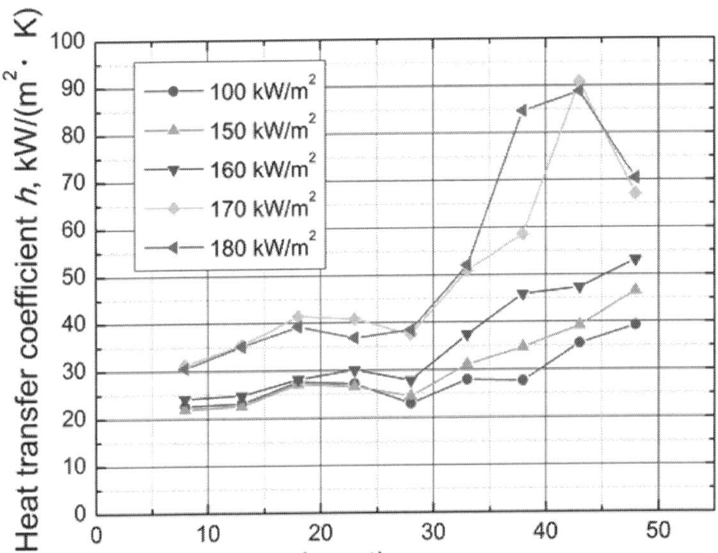

Figure 9. Heat transfer coefficient distribution on cryoprobe surface (Location z = 0 represents tip of cryoprobe).

5. MODE OF PHASE CHANGE HEAT TRANSFER IN CRYOPROBE

To evaluate the phase change heat transfer regime, the bubble expansion model [16] was applied to the ultrafine cryoprobe, as shown in Figure 11. Here, the bubble is assumed to travel at the velocity of the liquid phase and to expand only downstream. In addition, the annular channel of the ultrafine cryoprobe is assumed to be the equivalent circular channel by using the hydraulic diameter.

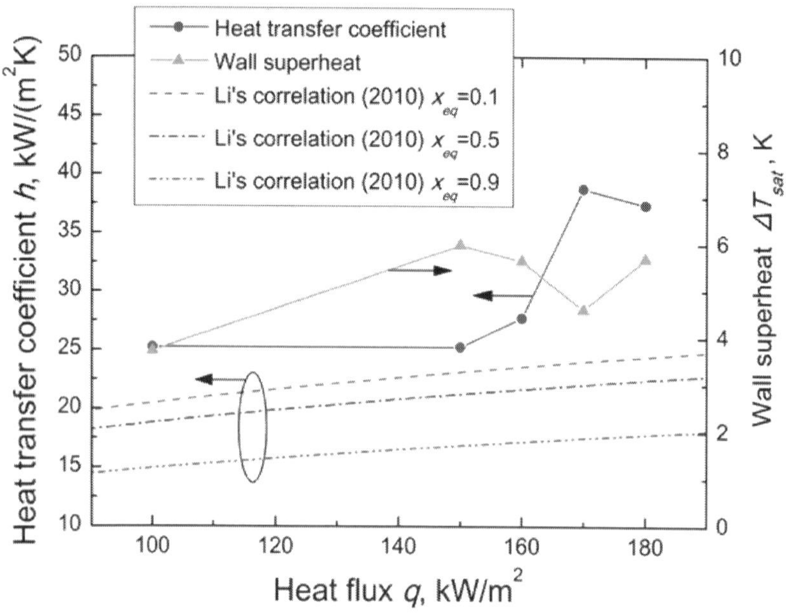

Figure 10. Relationship between heat transfer coefficient, wall superheat, and heat flux.

The energy balance between the bubble volume and supplied heat is expressed as

$$\rho_V i_{LV} \frac{\mathrm{d}V_B(t)}{\mathrm{d}t} = \pi D L_B(t) q \quad (8)$$

where V_B: bubble volume [m³], D: channel diameter [m], L_B: bubble length [m]. The bubble head speed U_B is defined as the time derivative of the bubble length as follows:

$$U_B(t) = \frac{\mathrm{d}L_B(t)}{\mathrm{d}t} \quad (9)$$

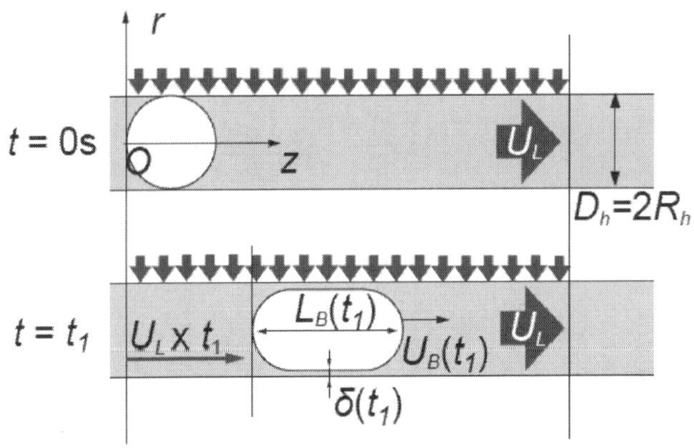

Figure 11. Bubble expansion model.

The liquid film thickness is determined by Taylor's law, which is the correlation between the liquid film thickness and bubble velocity, expressed as

$$\frac{2\delta}{D} = \frac{0.643(3Ca)^{2/3}}{1+1.608(3Ca)^{2/3}} \quad (10)$$

where δ : liquid film thickness [m], $Ca = \mu_L U_B / \sigma$: capillary number. This model can calculate the bubble expansion process and the temporal variation in the liquid film thickness during expansion. The heat transfer coefficient can be calculated from the liquid film thickness with following equation as $h = k_L / \delta$.

Figure 12 compares the experimentally obtained heat transfer coefficient with the theoretical prediction of the bubble expansion model. The theoretical prediction underestimates the experimental value. Additionally, Figure 13 compares the wall superheat from the experiment with the theoretical prediction of the bubble expansion model. The experimental results are much smaller than the theoretical prediction. In addition, the experimental results are larger than the minimum wall superheat for the onset of nucleate boiling [16]. Nucleate boiling does not occur if the wall superheat is less than this minimum value. Actually, it is difficult to measure the size and distribution of the cavity. Therefore, the minimum wall superheat to produce nucleate boiling is an important parameter. This parameter is determined by the relationship between the wall heat flux q_w and the wall superheat ΔT_{sat} to initiate heterogeneous nucleation, which is expressed as [17]

$$\Delta T_{sat} = \frac{\frac{R_g T_{sat}^2}{i_{LV}} \ln\left(1 + \frac{2\sigma}{p_L r_a}\right)}{1 - \frac{R_g T_{sat}}{i_{LV}} \ln\left(1 + \frac{2\sigma}{p_L r_a}\right)} + \frac{q r_a}{k_L}, \tag{11}$$

where R_g : gas constatnt [J/(kg・K)], r_a : active cavity size [m], and p_L : pressure in the liquid phase [Pa]. Therefore, the liquid film thickness predicted by the bubble expansion model is too thick to maintain a low wall temperature. Hence, the heat transfer mode in the ultrafine cryoprobe is not pure evaporative heat transfer from the liquid film. These results may indicate that phase change heat transfer in the ultrafine cryoprobe occurs by nucleate boiling.

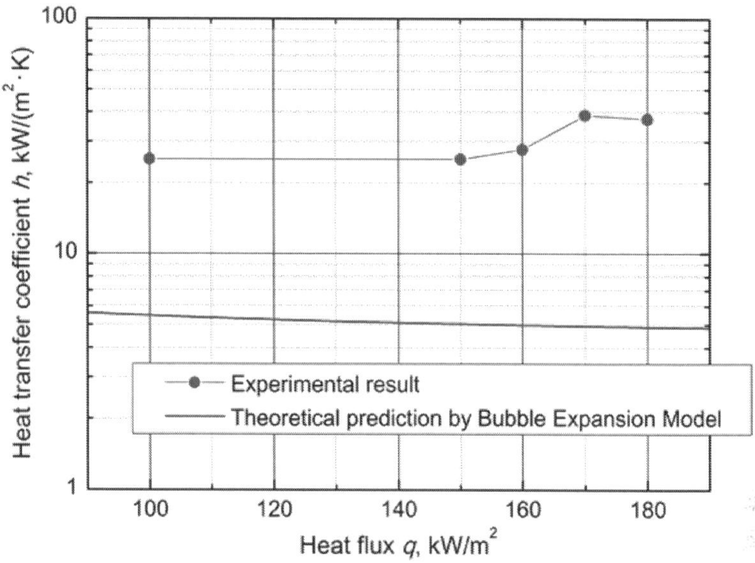

Figure 12. Comparison of heat transfer coefficients estimated from experimental data and from bubble expansion model.

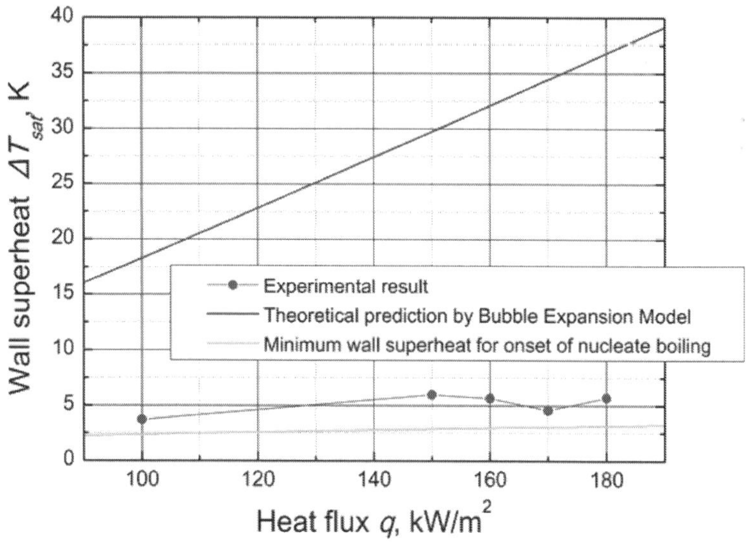

Figure 13. Comparison of wall superheat estimated from experimental data and that estimated from bubble expansion model.

6. CONCLUSIONS

In this study, the characteristics of phase change heat transfer in an ultrafine cryoprobe were evaluated. The heat transfer coefficient obtained from the experimental data was compared with that obtained from the conventional correlation and the bubble expansion model. The results obtained in this study can be summarized as follows:

1) The ultrafine cryoprobe cooled by HFC-23 exhibited a stable surface temperature of −50°C.

2) The heat transfer coefficient of the ultrafine cryoprobe was estimated. It reached a maximum value of 39 kW/(m^2 · K) at a heat flux of 170 kW/m^2.

3) By using the bubble expansion model, the phase change heat transfer regime in the ultrafine cryoprobe was estimated. The results may indicate that nucleate boiling occurs in the cryoprobe.

By using the developed cryoprobe, the small-scale cryosurgery will be achieved. We can take advantage of the small size of cryoprobe to treat small lesions, such as pigments on the skin, wrinkles around the eyes, and early breast cancer, with minimal invasiveness. Additionally, the cryosurgery in the blood vessel combined with the catheter is a possible application in the future.

ACKNOWLEDGEMENTS

Junnosuke Okajima, Atsuki Komiya, Shigenao Maruyama (B) [25820054] from the Japan Society for the Promotion of Science.

REFERENCES

1. Bischof, J., Christov, K. and Rubinsky, B. (1993) A Morphological-Study of Cooling Rate Response in Normal and Neoplastic Human Liver-Tissue-Cryosurgical Implications. Cryobiology, 30, 482-492.

2. Seifert, J.K., Gerharz, C.D., Mattes, F., Nassir, F., Fachinger, K., Beil, C. and Junginger, T. (2003) A Pig Model of Hepatic Cryotherapy. In Vivo Temperature Distribution during Freezing and Histopathological Changes. Cryobiology, 47, 214-226.

3. Rewcastle, J.C., Sandison, G.A., Saliken, J.C., Donnelly, B.J. and McKinnon, J.G. (1999) Considerations during Clinical Operation of Two Commercially Available Cryomachines. Journal of Surgical Oncology, 71, 106-111.

4. Popken, F., Land, M., Bosse, M., Erberich, H., Meschede, P., Konig, D.P., Fischer, J.H. and Eysel, P. (2003) Cryosurgery in Long Bones with New Miniature Cryoprobe: An Experimental in Vivo Study of the Cryosurgical Temperature Field in Sheep. European Journal of Surgical Oncology, 29, 542-547.

5. Tacke, J., Adam, G., Haage, P., Sellhaus, B., Großkortenhaus, S. and Günther, R.W. (2001) MR-Guided Percutaneous Cryotherapy of the Liver: In Vivo Evaluation with Histologic Correlation in an Animal Model. Journal of Magnetic Resonance Imaging, 13, 50-56.

6. Doll, N., Meyer, R., Walther, T. and Mohr, F.W. (2004) A New Cryoprobe for Intraoperative Ablation of Atrial Fibrillation. The Annals of Thoracic Surgery, 77, 1460-1462.

7. Takeda, H., Maruyama, S., Okajima, J., Aiba, S. and Komiya, A. (2009) Development and Estimation of a Novel Cryoprobe Utilizing the Peltier Effect for Precise and Safe Cryosurgery. Cryobiology, 59, 275-284.

8. Aihara, T., Kim, J.-K., Suzuki, K. and Kasahara, K. (1993) Boiling Heat Transfer of a Micro-Impinging Jet of Liquid Nitrogen in a Very Slender Cryoprobe. International Journal of Heat and Mass Transfer, 36, 169-175.

9. Maruyama, S., Nakagawa, K., Takeda, H., Aiba, S. and Komiya, A. (2008) The Flexible Cryoprobe Using Peltier Effect for Heat Transfer Control. Journal of Biomechanical Science and Engineering, 3, 138-150.

10. Bénita, M. and Condé, H. (1972) Effects of Local Cooling upon Conduction and Synaptic Transmission. Brain Research, 36, 133-151.

11. Zhang, J.-X., Ni, H. and Harper, R.M. (1986) A Miniaturized Cryoprobe for Functional Neuronal Blockade in Freely Moving Animals. Journal of Neuroscience Methods, 16, 79-87.

12. Okajima, J., Komiya, A. and Maruyama, S. (2010) Boiling Heat Transfer in Small Channel for Development of Ultrafine Cryoprobe. International Journal of Heat and Fluid Flow, 31, 1012-1018.

13. Okajima, J., Maruyama, S., Takeda, H., Komiya, A. and Sangkwon, J. (2010) Cooling Characteristics of Ultrafine Cryoprobe Utilizing Convective Boiling Heat Transfer in Microchannel. Proceedings of 14th International Heat Transfer Conference, Washington DC, 8-13 August 2010, 297-306.

14. Okajima, J., Komiya, A. and Maruyama, S. (2013) 24-Gauge Ultrafine Cryoprobe with Diameter of 550 μm and Its Cooling Performance. Submitted to Cryobiology.

15. Li, W. and Wu, Z. (2010) A General Correlation for Evaporative Heat Transfer in Micro/Mini-Channels. International Journal of Heat and Mass Transfer, 53, 1778-1787.

16. Okajima, J., Komiya, A. and Maruyama, S. (2012) Analysis of Evaporative Heat Transfer by Expansion Bubble in a Microchannel for High Heat Flux Cooling. Journal of Thermal Science and Technology, 7, 740-752.

17. Davis, E.J. and Anderson, G.H. (1966) The Incipience of Nucleate Boiling in Forced Convection Flow. AIChE Journal, 12, 774-780.

4

HEAT TRANSFER ANALYSIS DURING EXTERNAL CHILLING OF COMPOSITE MATERIAL CASTINGS THROUGH EXPERIMENTAL AND FINITE ELEMENT (FE) MODELLING

Joel Hemanth

Mechanical Engineering, H.M.S. Institute of Technology,
Tumkur, India

ABSTRACT

In controlled solidification, one of the important factors that affects heat transfer from the solidifying casting is the resistance offered at the casting/chill interface. In the present investigation, heat transfer analysis during solidification of Al-12%Si (LM 13) alloy is carried out by collecting temperature history of the solidifying casting. The temperature distribution during solidification in the present investigation is obtained using ANSYS multi physics software and further for comparison. The temperature profiles are also obtained by FE (Finite Element) modelling using the same software. By using a temperature data logger and lab view based software, the temperature data is acquired and processed at every second. The cooling curves obtained are analysed to know the effect of chilling on solidification behaviour of Al-12%Si alloy castings. Finally, it is concluded from the above research that the cooling curves and temperature distribution obtained by FE analysis do not so closely converge with the experimental data due to modelling limitations.

KEYWORDS

Solidification; Heat; Chill; Casting; Temperature

1. INTRODUCTION

With the increasing demand for stringent properties and high quality castings, controlled solidification has become the present need. Many of the early metallic materials failed when exposed to complex requirements of industrial, automotive and structural applications due to unsoundness.

Hence, the controlled heat transfer during solidification forms an essential area of investigation. Controlling the rate of solidification using chills is one among many existing foundry practices and use of chills during the solidification plays a major role in promoting directional solidification [1-5].

In recent years, research in the area of numerical modelling of solidification is gaining much importance [6].

The computer simulation of freezing patterns in casting has done much to broaden the understanding of solidification of materials. The way by which the heat flows across the casting/chill interface strongly affects the evaluation of solidification and plays a remarkable role in the structural integrity of castings. Heat transfer at chill/metal interface during solidification is the driving factor in determining the microstructure and mechanical properties of the material [7,8]. It becomes significant when the solidifying metal and the chill have reasonably good conductance. Depending on the material and the size, chills increase heat transfer rate affecting the microstructure and hence altering the mechanical properties of the castings. To understand the behaviour at chill/casting interface during solidification, several experimental methods are available [9] since solidification does not occur under equilibrium conditions. But still the cooling curve method is popularly used to understand and interpret the temperature-time data collected during solidification [10-13].

It is often difficult to obtain the analytical solution for phase change problems in multi-dimensions. Therefore, in the literature, some preliminary studies of solidification of metals using a finite difference technique have been reported [14] in order to analyse the factors affecting the rate of cooling. These factors include the material properties, and the heat sink temperature. The previous results, however, have neglected the effect of natural convection in the liquid melt [15]. Several experimental results have confirmed that convection has an important role in solid/liquid phase change [16-18].

The simulation method was based on an axi-symmetric model with enthalpy formulation of the energy equation coupled with the momentum equations and thus simulation has taken natural convection into account during the freezing process. Further, calculations are done by using the finite volume analysis with SIMPLER algorithm for pressure correction and other processing parameters affecting the rate of solidification [19].

Earlier, there have been several attempts to model and simulate the process of solidification of materials but the physics of microstructural solidification is not completely understood [20]. In addition, the generation of accurate finite element and finite difference solutions may take prohibitive amounts of computational time. Brown and Spittle [21] pointed out from their investigation that finite element and finite difference analysis methods can offer powerful solutions to solidification problems. The primary and most obvious phenomenon during solidification is the transfer of heat from the cooling casting to the sand mould. Provision of chills in the mol corresponds to an appropriate interfacial heat-transfer coefficient value across the chill/metal interface [22-25]. Since the objective of feeding design is to keep hot spots in the feeder, the reliability of a numerical analysis would largely depend on the heat-transfer modelling across the metal-mould interface. There has been great interest on the experimental side as well as on the numerical research side in understanding the complexity of the heat transfer across the casting mould interface. In the present investigation, heat transfer at the chill/casting interface is obtained by obtaining temperature history at the interface using thermocouples. Thermocouples are extensively used to measure dynamic or transient temperature history at strategic locations in the mould or in the solidifying melt.

The temperature history is obtained using data logging and acquisition system. The cooling curve method is the most widely used method for investigating solidification of metals and alloys. Cooling curves thus obtained from the data are finally analysed. In order to know the temperature distribution during solidification, a simple macro modelling and simulation of casting was carried out using a software code. Therefore, the present work uses ANSYS multi physics code to model phase change during solidification and to know the temperature distribution during solidification. The application and setup is the most simplistic and therefore can be widely used by both academic and industry researchers. The setup involves placing a thermocouple in the melt and allowing the melt to solidify while temperature is recorded as a function of time.

2. EXPERIMENTAL PROCEDURE

The heat transfer analysis during solidification in the present investigation is carried out by thermo-couple method. Encapsulated K-type (Chromel-alumel) thermocouples were inserted in to sand mould at location T1, the chill/casting interface, and at other two locations T2 & T3, entirely within the solidifying casting. The locations on the centre line are at chill/casting interface (T1) and at other two points at 20 mm (T2) and 40 mm (T3) away from it (**Figure 1**). The thermocouples are connected to the temperature data logger, which is inturn, connected to the computer. Using Lab view based Scanner IV software, the temperature data is acquired and processed every second after the melt is poured into the sand mould at 700°C. Pilot trials are run and all steps are taken to avoid uncertainties during temperature measurement. **Figure 2** shows Al-12% Si alloy castings after solidification along with thermocouples.

The casting dimensions of 112.5 mm × 75 mm × 25 mm (AFS standard) are considered for heat transfer analysis. The temperature history is acquired for 20 minutes and is processed by the software to give temperature Vs time graph (cooling curve). The analysis is carried out for solidification against copper end chill. The temperature data logger is capable of collecting data from 12 channels at a time and the unit is provided with LCD display for temperature, channel number and scan delay. The data from the temperature data logger is fed to PC through RS232 communication port where the data is captured through data acquisition software.

2.1. Finite Element Modelling and Meshing

The 2-D analysis of the composite casting is carried out assuming unit thickness. The geometry of the model is as shown in the **Figure 3** and the finite element model is as shown in the **Figure 4**. Finite elements model consist of 8435 2-D thermal solid elements, 6935 for the sand mould, 992 for the metal and 400 for the chill. The total number of nodes in the model is 8035.

The element has four nodes with a single degree of freedom (temperature) at each node.

The shape function for this finite element is:

$$T = \frac{1}{4}\left(T_I(1-s)(1-t) + T_J(1+s)(1-t) + T_K(1+s)(1+t) + T_L(1-s)(1+t)\right) \quad (1)$$

where T_I, T_J, T_K, and T_L are nodal temperatures, s and t axis of local (element based) coordinate system. Boundary and initial conditions: on the boundary of the model, excluding axis of symmetry, a constant value for film coefficient (hf = 7.5W/m^2°C), air temperature (T_0 = 30°C), uniformly distributed initial temperatures of sand (T_0 = 30°C) and alloy (T_2 = 700°C) temperatures were considered.

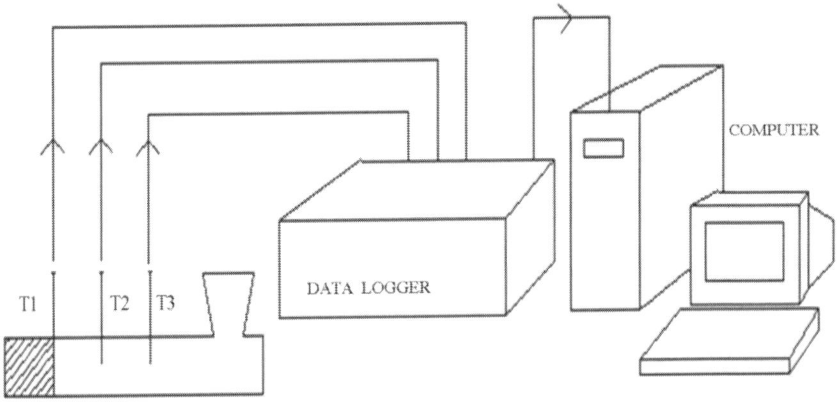

Figure 1. Schematic representation of the experimental setup with data acquisition system.

2.2. Finite Element Procedure

The analysis is carried out for copper end at temperatures T1, T2, & T3 locations of the casting during solidification and the cooling curves are also obtained for the analysis. The geometry of the FE model is as shown in **Figure 5**.

Figure 2. Castings after solidification along with thermocouples.

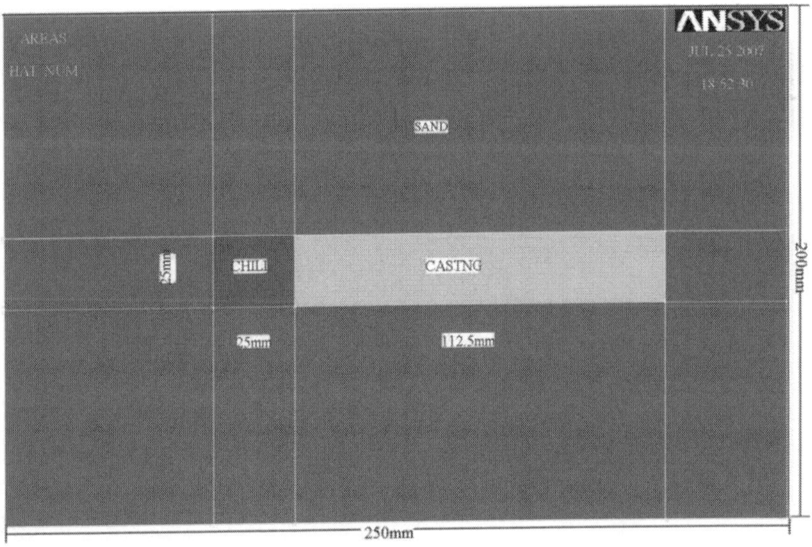

Figure 3. Geometry of Finite Element (FE) model.

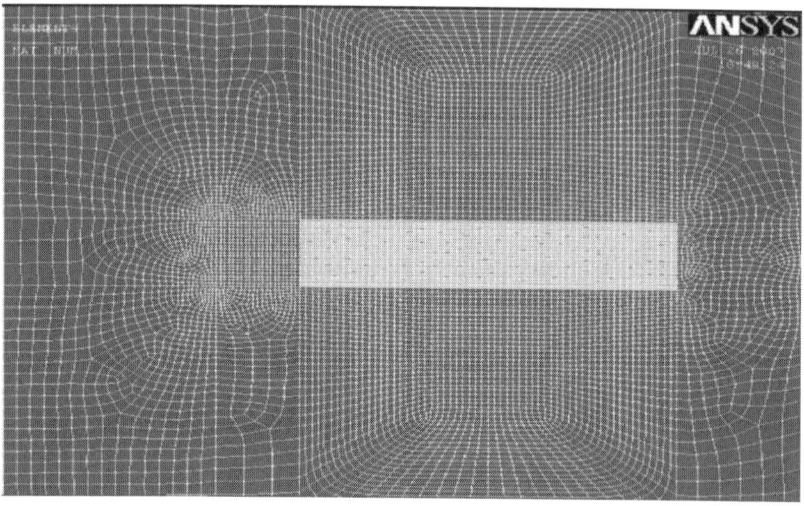

Figure 4. Meshed model of the chill and the mould.

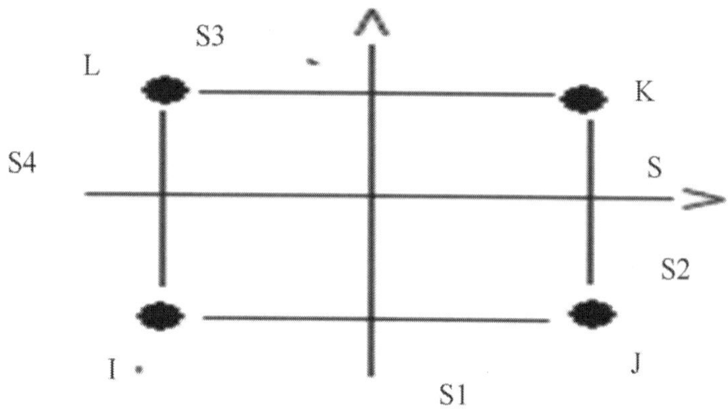

Figure 5. Geometry of the finite element model.

The part to be cast is assumed to be at the centre of the mould box of length 250 mm and width 200 mm. The molten metal is introduced into the mould at 700°C. The mould is initially at 30°C. The top face and side of the mould exchange heat with the environment by free convection. The ambient temperature is taken as 30°C and film coefficient on the side of the mould is 7.5°C W/m²°C and 5.75 W/m²°C on the top. Axi-symmetric behaviour is assumed for sand mould and metal melt.

The temperatures of the melding sand and the chill are assumed to be constant and uniform throughout. A transient analysis is set up using initial conditions with sand mould at 30°C and metal melt at 700°C. The side and top convective loads are included on the mould exterior. No surface boundary conditions are specified on the bottom or centreline.

In order to understand temperature distribution during solidification heat transfer, two-dimensional FE modelling and analysis has been investigated. The comercially available ANSYS multi-physics software code is used for the present analysis. For simplicity, only conductive heat transfer after pouring is considered for modelling and analysis but physical properties of chill and the sand materials do not vary during solidification. Varying specific heat and thermal conductivity are considered as material properties. Modelling is done using PLANE55 first order quadrilateral element. The ANSYS code uses backward Euler time integration method for solution [26].

3. RESULTS, DISCUSSIONS AND ANALYSIS

3.1. Numerical Modelling

According to the first law of thermodynamics, the total heat of a closed system remains unchanged. As applied on differential control volume, one can state:

$$\rho c \partial T / \partial t + \rho c \{v\}^T \{L\} T + \{L\}^T \{q\} = \bar{q} \quad \ldots\ldots \quad (2)$$

where: ρ, density, c, specific heat, t, time, $T = T(x,y,z,t)$, temperature distribution function, $\{q\}$ - heat flux vector, $\bar{q}$ -heat generation rate per unit volume. Equation (2) is the so-called transport equation for thermal energy.

It should be realized that the terms $\{L\}T$ and $\{L\}T\{q\}$ might also be interpreted as grad T and $\mathrm{div}(\{q\})$, respectively. Then, it clearly highlights the various transport processes: the first term on the left hand side of Equation (2) is rate of change term, the second is convective term, and the third is diffusive term, while on the right hand side of the equation there is the source term (regarding a differential control volume). Fourier's law was used to relate the heat flux vector to the thermal gradients:

$$\{q\} = -[D]\{L\}T \cdots \quad (3)$$

[D] is the conductivity matrix, Kxx, Kyy, Kzz-conductivity in the element x, y, and z directions, respectively.

Combining Equations (2) and (3), assuming that all effects are related to the global Cartesian system:

$$\begin{aligned} &\rho c\left(\partial T/\partial t + v_x \partial T/\partial x + v_y \partial T/\partial y + v_z \partial T/\partial z\right) \\ &= Q + \partial/\partial x\left(K_x \partial T/\partial x\right) \\ &\quad + \partial/\partial y\left(K_y \partial T/\partial y\right) + \partial/\partial z\left(K_z \partial T/\partial z\right) \end{aligned} \quad (4)$$

Three types of boundary conditions are considered. It presumes that these cover the entire element.

a) T = T*, where T* is specified temperatures acting over surface S1;

b) $\{q\}^T\{h\} = -q^*$, where q*-specified heat flow acting over surface S2, $\{\eta\}$ - unit outward normal vector, c) $\{q\}^T\{h\} = h_f(T_S - T_B)$, specified convection surfaces acting over surface S3 (Newton's law of cooling), where hf is film coefficient evaluated at $(T_B+T_S)/2$ (unless otherwise specified for the element), T_B-bulk temperature of the adjacent fluid, T_S-temperature at the surface of the model.

If we introduce $\{T_e\}=\{Te(t)\}$ as vector of nodal temperatures, changes in temperature over element and time domain $(T=T(x,y,z,t))$ could be presented as:

$$T=\{N\}^T\{T_e\} \tag{5}$$

where $\{N\}=\{N(x,y,z)\}$, element shape function.

After some manipulation and integration over element volume, using relation $\{L\}T=[B]\{T_e\}$, where $[B]=\{L\}\{N\}$ T Equation (4) can be written as:

$$\begin{aligned}
&\int_v \rho c\{\delta T_e\}^T\{N\}\{N\}^T\{T_e\}\mathrm{d}(V)\\
&\quad+\int_v \rho c\{\delta T_e\}^T\{N\}\{v\}^T[B]\{T_e\}\mathrm{d}(V)\\
&\quad+\int_v\{\delta T_e\}^T\{B\}^T[DIB]\{T_e\}\mathrm{d}(V)\\
&=\int_{S_1}\{\delta T_e\}^T\{N\}q^*\mathrm{d}(S_2)\\
&\quad+\int_{S_3}\{\delta T_e\}^T\{N\}h_f\left(T_B-\{N\}^T\{T_e\}\mathrm{d}(S_3)\right)\\
&\quad+\int_v\{\delta T_e\}^T\{N\}q\mathrm{d}(V)
\end{aligned} \tag{6}$$

Mould density assumed to remain constant over the volume of the element. Nodal quantities $\{Te\}$, $\{Te\}$ and $\{\delta Te\}$ are constant in the domain of the element and they can be removed from the integral, while c and q may vary over the element, so after some manipulation yields to the matrix form of law of heat energy conservation:

$$\begin{aligned}
&\left[C_e^t\right]\{Te\}+\left(\left[K_e^{tm}\right]+\left[K_e^{tb}\right]+\left[K_e^{tc}\right]\right)\{Te\}\\
&=\{Q_e^f\}+\{Q_e^c\}+\{Q_e^g\}
\end{aligned} \tag{7}$$

where,

$\left[C_e^t\right] = \rho\int_v C\{N\}\{N\}^T \mathrm{d}V$ -element specific heat (thermal damping) matrix.

For the phase change problems, members of this matrix are defined from the enthalpy vs. temperature curve, according enthalpy model, and taking into account latent heat of phase change.

$\left[K_e^{tm}\right] = \rho\int_v C\{N\}\{v\}^T [B]\mathrm{d}V$, element mass transport conductivity matrix;

$\left[K_e^{tb}\right] = \int_v [B]^T [D][B]\mathrm{d}V$, element diffusion conductivity matrix;

$\left[K_e^{tc}\right] = \int_{S_1} h_f \{N\}\{N\}^T \mathrm{d}(S_3)$, element convection surface conductivity matrix; this matrix is diagonal, with the diagonal terms defined by the vector $\{Q_e^f\} = \int_{S_2}\{N\}q^*\mathrm{d}(S_2)\ \int_{S_3} h_f \{N\} d(S_3)$, element mass flux vector.

$\{Q_e^c\} = \int_{S_3} T_B h_f \{N\}\mathrm{d}(S_3)$, element convection surface heat flow vector;

$\{Q_e^g\} = \int_V \ddot{q}\{N\}\mathrm{d}(V)$, element heat generation load.

3.2. Numerical Procedure

After defining conductivity matrices for each finite element as in Equation (7), one can obtain a set of linear equations defining behaviour of whole system of finite elements. This set has the following form:

$$[C]\{\dot{T}\} + [K]\{T\} = \{Q^a\}$$

where [C], specific heat matrix of the system; [K], conductivity matrix of the system; {T}, nodal temperature vector; $\{Q^a\}$, external heat load.

While solving phase change problem, the above transient heat conduction equation is solved. The second and the term on to the right of the above equation is not affected by phase change. The phase change is accounted by $[C]=\rho\int_{v} C\{N\}\{N\}^{T} dV$ in the above equation. This system is solved using full Newton-Ralphson method and an Euler backward difference scheme is used for the transient time integration. In order to avoid numerical instability, the maximum time step size was controlled by the description of the loading input and the minimum time step size (or maximum element size) by defining its value based on the following relationship: $ITS = \Delta^2/4\alpha$.

3.3. MNSMS

The Δ value is the conducting length of an element (along the direction of heat flow) in the expected highest temperature gradient. The α value is the thermal diffusivity, given by k/ρC. The k value is the thermal conductivity, ρ is the mass density, and C is the specific heat. If the above relationship (ITS = $\Delta^2/4\alpha$) is violated, unwanted numerical oscillations and values of calculated temperature outside of the physically possible range, may occur [27].

3.4. Cooling Curves during Solidification of Al-12%Si Alloy Cast against a Copper Chill

The cooling curves at T1, T2 & T3 locations inside casting during solidification of Al-12%Si alloy are as shown in the **Figure 6**. The solidification at T1 location has begun at a lower temperature compared to T2 & T3 locations due to the presence of copper chill. The difference in temperature between T1 and T3 locations is around 50°C. This decrease is because of both chilling and convective heat transfer in the melt. The curve portion in the mushy region for T1 & T2 locations is steeper than the curve at T3 location. This signifies that the liquid-solid transformation takes place much earlier near the chill end than at locations away from it. The decrease in temperature throughout solidification at T3 location maintains temperature difference from T1 & T2 locations. Whereas the curve at T1 seems to rise above T2 curve during cooling phase of solid-solid transformation. This is because of chill temperature which rises during solidification does not decline at the same rate.

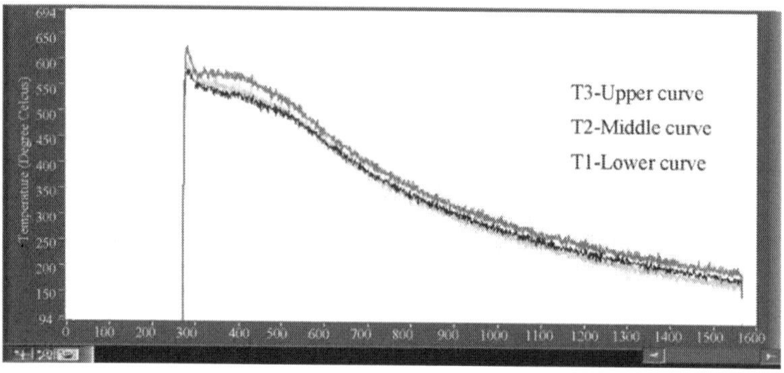

Figure 6. Cooling curves (temperature in °C verses time in seconds) obtained at thermocouple locations T1, T2 and T3 for Al-12%Si alloy cooling against a copper chill.

The presence of chill has influenced the solidification behaviour as shown in Figures 6. The solidification curves at T1, T2 and T3 locations of Al-12%Si alloy cooling against copper chill showed a distinct behaviour because of super cooling at the melt/chill interface. This indicates the effect of chill/melt interfacial heat transfer that has affected the solidification behaviour. The onset of nucleation is at lower temperature near chill/melt (T1) interface compared to locations at T2 and T3. To reduce air gap resistance, the chill surfaces facing the composite melt are given very fine finish. Observably, the solidification in the mushy region at the chill/melt interface (T1) is eutectic type compared to the locations at T2 and T3. After three minutes, the cooling rate for all the three locations has been stabilized. The cooling curves thus shows that in the last phase of solidification heat transfer, the T1 curve has shown upward trend compared to T2 curve. During the last phase of solidification (solid-solid cooling), the melt at T1 location is affected by the cooling rate of the hot chill. Since cooling rate of the hot chill differs from other parts of the casting, the T1 curve shows higher temperature than T2 curve during last stages of solidification.

3.5. Temperature Profiles for Al-12%Si Alloy Cast against a Copper Chill

The predicted temperature profiles of Al-12%Si alloy solidifying against copper chill at 10 sec., 75 sec. and 180 sec. are shown in the Figures 7-9. It is found from the temperature profile that, heat transfer has taken place towards the chill end from the riser end with mushy region at 75 sec. and completion of solidification at 180 sec. The temperature profile of the solidifying alloy at different times and temperature at chill/metal interface and the riser end are generated by the program. Finally it is found that at the end of 75 seconds, maximum melt temperature at riser end was 405°C and at chill end 361°C.

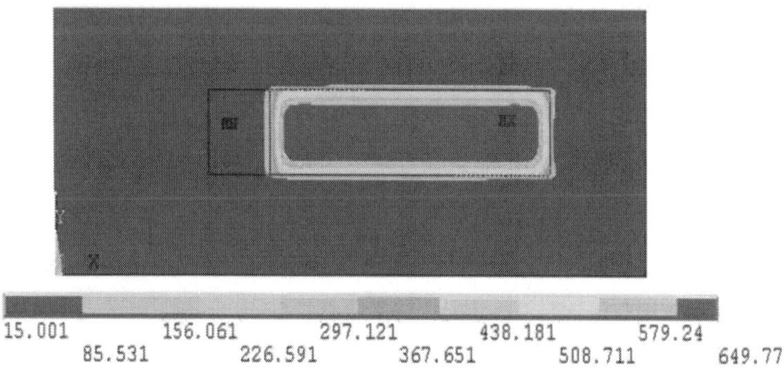

Figure 7. Predicted temperature profiles of solidifying base alloy (Al-12%Si) under the influence of copper chill at the end of 10 second.

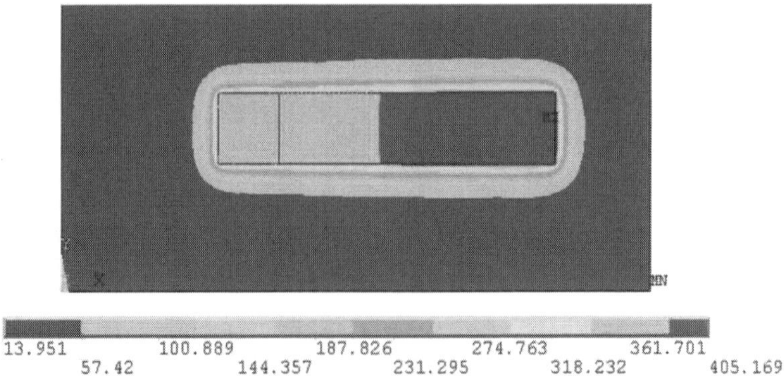

Figure 8. Predicted temperature profiles of solidifying base alloy (Al-12%Si) under the influence of copper chill at the end of 75 seconds.

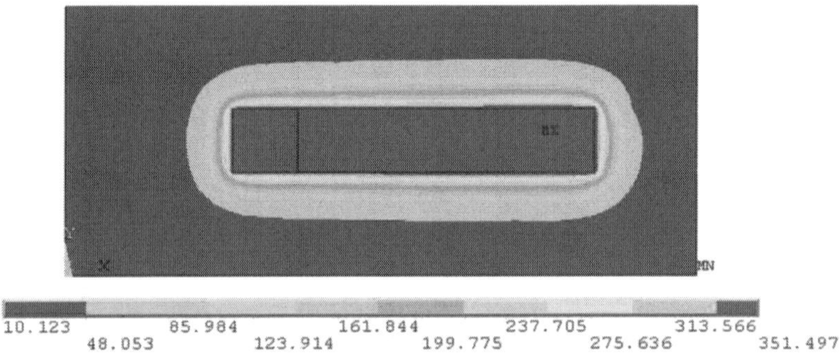

Figure 9. Predicted temperature profiles of solidifying base alloy (Al-12%Si) under the influence of copper chill at the end of 180 seconds.

The results shown by the temperature profiles of the solidifying are also generated by the program. In general, the profile exhibit a temperature gradient that reduces towards the chill end. This indicates that the chill is establishing the directional solidification and heat transfer is taking place from riser end to chill end. Hence chills has influenced appreciably the temperature distribution along the length of the casting. In the first three minutes after pouring, the copper chilled Al-12%Si alloy has shown higher temperatures at 10 sec. and reduces after 75 sec. and 180 sec. Observation also indicated that the temperature profile results of FE analysis is qualitatively in line with the experimental cooling curves. It can be seen from the experimental cooling curves that the dip in the initial temperature for the Al alloy when chilled is observably more. This fact is further evidenced by the cooling curves obtained by the FE analysis. T his shows that chill material has affected the heat transfer and hence the temperature distributions in the solidifying melt. However, the overall cooling rates are controlled by heat transfer at chill end, heat transfer through sand, metallurgical condition of the melt and convective currents.

3.6. Cooling Curve from FE Analysis

The cooling curves obtained by FE simulation of the solidifying Al-12%Si alloy cast against copper end-chill. The cooling curve is obtained at three locations inside the solidifying melt such as chill/melt interface (T1), 20 mm from chill (T2) and 40 mm from chill (T3). These locations are considered along the centre line of the casting thickness (25 mm). It is observed that in the first three minutes, the copper chilled Al-12%Si alloy has solidified at relatively higher temperatures at T1, T2 & T3 locations.

The present FE analysis assumes only conductive heat transfer during solidification whereas convective heat transfer in the melt during and after pouring also exists. This is because modelling of solidification phenomena becomes very complex if convection heat transfer is also taken into account. Hence it is observed that there is a little mismatch which exist between the experimental and FE cooling curves due to modelling limitations. With these modelling limitations, the FEM cooling curves cannot be quantitatively compared with the experimental results.

4. CONCLUSIONS

The cooling curves obtained at chill/melt interface and at different locations of the solidifying melt indicate that the chills have a pronounced effect on the solidification. The solidification at chill/melt interface starts at lower temperatures compared to other locations along the length of the casting. This indicates that the chills have established steep temperature gradient along the length of casting and promote directional solidification.

A successful attempt has been made to model phase change and to know temperature distribution in composites during solidification using ANSYS-10 FE comercial software.

With the modelling limitations, the results give some insights into temperature distribution during Al-12%Si alloy solidification. Observing the temperature gradients obtained, conclusion can be drawn that the heat transfer has occurred from riser end to chill end promoting directional solidification. Within three minutes of solidification, the cooling curves obtained by FEM model shows similar trend as that of experimental curves. It can be seen from the experimental cooling curves that the dip in the initial temperature for the Al alloy when chilled is observably more. Finally, the FE cooling curves cannot be compared fully and quantitatively with experimental cooling curves because of modelling limitations.

REFERENCES

1. J. Hemanth, "Fracture Behaviour of Cryogenically Solidified Al-Alloy Reinforced with Nano ZrO2MMCs," International Journal of Microstructure and Materials Properties (IJMMP), Vol. 7, No. 4, 2012, pp. 68-77.

2. J. Hemanth, "Microstructure, Mechanical Properties and Wear Behaviour of Metallic, Non-Metallic and Deep Cryogenically

Chilled ASTM a216 WCB Steel," Journal of Alloys and Compounds, Elsevier Science, Vol. 506, 2010, pp. 645-652.

3. J. Hemanth, "Development and Wear Behaviour of Al/ Al_2SiO_5/C Chilled Hybrid Metal Matrix Composites by Both Experimental and Finite Element Method," SAE International, Ohio, 2011.

4. J. Hemanth, "Abrasive and Slurry Wear Behaviour of Chilled Aluminium Alloy (A356) Reinforced with Used Silica (SiO_2p) MMCs," Composites Part B, Elsevier Science, Vol. 42, No. 7, 2011, pp. 1826-1833.

5. J. Hemanth, "Cryo Effect during Solidification on the Tribological Behaviour of Al-Alloy/GlassSiO_2 MMCs," Journal of Composite Materials, Vol. 43, No. 6, 2009, pp. 675-688.

6. P. K. Rohatgi, et al., "Heat Transfer Characteristics of Mould Materials," Journal of Materials Science, Vol. 29, No. 20, 1994, pp. 5357-5366.

7. G. S. Hanumanth and G. A. Irons, "Solidification Behaviour of Aluminium Alloy Die Castings," Metallurgical and Materials Transactions B, Vol. 27, No. 4, 1996, pp. 663-671.

8. S. L. Soo, "Multiphase Fluid Dynamics," Science Press, Beijing, 1990.

9. G. S. Hanumanth, G. A. Irons and S. Lafreniere, "Effect of Chill Metal Interface during Solidification of Aluminium Alloys," Metallurgical and Materials Transactions B, Vol. 23, No. 6, 1992, pp. 753-763.

10. C. Beckermann and C. Y. Wang, "Annual Review of Heat Transfer," In: C. L. Tien, Ed., Begell House, New York, 1995, pp. 98-115.

11. A. Scott, "Cooling Curve Analysis of Chilled Aluminium 12% Si Alloys," M.S. Thesis, Virginia Polytechnic Institute and State University, Blacksburg, 1996.

12. K. N. Seetharamu and Sadhana, "Solidification Behaviour of Aluminiumalloy Composite by Die Casting Method," Journal of Composites, Vol. 26, 2009, pp. 103-120.

13. R. W. Lewis and R. S. Ransingh, "Solidification Analysis through Cooling Curves for Aluminium Alloy Reinforced with SiO_2 Composites," Metallurgical and Materials Transactions, Vol. 29B, 1988, pp. 435-444.

14. K. S. Kannan, et al., "Effect of Heat Transfer on during Solidification of Steel Castings," Indian Journal of Technology, Vol. 28, 1990, pp. 460-474.

15. L. Bäckerud and B. Chalmers, "Effect of Convection Heat Transfer during Solidification of Aluminium Alloy Castings," Transactions of the Metallurgical Society of AIME, Vol. 245, No. 2, 1969, pp. 309-318.

16. L.-S. Chao and W.-C. Du, "Effect of Air Gap Formation in Chills Ruing Solidification of Copper Alloy Castings," Proc. Natl. Sci. Counc. ROC (A), Vol. 23, No. 5, 1999, pp. 622-629.

17. J. Tamminen, "Heat Transfer Analysis between Chill and Metal Interface," Ph.D. Thesis, University of Stockholm, Stockholm, 2010.

18. H. Morrogh and W. J. Williams, "Convective Heat Transfer Analysis during Solidification of Steel Alloys," Journal of Iron and Steel Institute, Vol. 12, 1999, pp. 375- 378.

19. A. Nathan, "Correction of Transient Solid-Embedded Thermocouple Data with Application to Inverse Heat Conduction," Ph.D. Thesis, Mississippi State University, Mississippi, 2011.

20. ANSYS Software Company, "Manual of ANSYS-10 Multi-Physics Tutorials on Phase Change Problems," 2011.

21. C. A. Long, "Essential Heat transfer," Addition Wesley Longman Publications, Boston, 2001.

22. R. Asthana, "Finite Element Analysis of Solidification Behaviour of Metals," Journal of Materials Science, Vol. 15, No. 11, 1998, pp. 213-255.

23. J. Hashim, L. Looney and M. S. J. Hashmi, "Finite Element Modelling of Heat Transfer Analysis during Solidification of Aluminium Alloy Castings," Journal of Material Processing Technology, Vol. 123, No. 2, 2002, pp. 251-257.

24. Z. Markovie and B. Rasuo, "Finite Element and Finite Difference Method of Analysing Hear Transfer during Solidification," International Conference on Computational Mechanics (CM04), Belgrade, November 2004, pp. 26-34.

25. T. Goel and M. Rowan, "Film Coefficient of Heat Transfer for Aluminium Alloy Chilled Castings," Material Science Letters, Elsevier Science, Vol 25, 2010, pp. 81- 89.

26. M. Hughes and K. L. Warren, "ANSYS Code for Euler's Backward Integration Method for Solution to Heat Transfer during Solidification Problems," Journal of Advanced Heat Transfer, Vol. 17, No. 2, 2011, pp. 34-47.

27. S. Sawato, J. L. Guy and L. H. Philips, "Temperature Gradients during Solidification—A Numerical Approach," Journal of AIME, Vol. 24, No. 9, 2010, pp. 56-67.

5

MODELING AND SIMULATION OF LASER ASSISTED TURNING OF HARD STEELS

Omar Abdulghani[1,2], Mohamed Sobih[3], Amro Youssef[3], Abdel-Monem El-Batahgy[4*]

[1]Military Technical College, Cairo, Egypt

[2]Libyan National Army, Tripoli, Libya

[3]Military Technical College, Cairo, Egypt

[4]Central Metallurgical R & D Institute, Cairo, Egypt

ABSTRACT

This research work is focused on simulation of laser assisted turning as a new solution for machining of hard steels. A transient, three-dimensional model was developed to predict the temperature distribution of a rotated cylindrical steel workpiece subjected to a localized heating using a moving Gaussian laser beam. In this regard, a User-Defined Function was created to overcome the problem of a moving Gaussian heat source' definition. This User-Defined Function was compiled into a finite volume software package (Fluent®), where three-dimensional single precision solver was used for analysis. Based on this model, simulation of the surface temperature of 32 mm diameter workpiece of AISI51 50H steel was performed as a function of time at a specific distance behind the laser beam spot, which is corresponding to 30° angle from the laser beam. The simulation results were compared with other published data of the same steel type where a close agreement was obtained. The verified model was used for simulation of laser assisted turning of 20 mm diameter workpiece of AISI D2 tool steel. The cutting depth, behind the laser beam, was set at a distance corresponding to 60° angle from the laser beam for having sufficient access for handling both laser head and cutting tool. This cutting depth was studied as a function of different lasers and machining parameters. The results indicated that the optimum parameters for successful laser-assisted turning process of the concerned steels are 800 W laser power, 5 mm laser beam spot diameter, 20 sec preheating time, 0.8 mm/sec laser scanning speed, 300 rpm rotational speed and 0.8 mm/sec feed rate. These parameters ensure easy/successful cutting of 1 mm depth in one pass without deteriorating the properties of the remaining bulk material. It can be deduced that the developed model might provide a useful tool for online process control of different steel types regardless of their physical properties and geometries.

KEYWORDS

Laser assisted turning; Hard steels; Three-dimensional modeling; User-defined function; Temperature distribution

1. INTRODUCTION

Hard steel alloys are widely used in various important applications such as automotive industry, die and mold industries as well as cutting purposes including shear blades, punches and piercers. Manufacturers of hard components constantly strive for lower cost solutions as well as higher quality in order to maintain their competitiveness. Conventional hard machining processes that may be used in pre-grinding or pre-finishing operations result in low quality. Besides, it is still an expensive and complicated solution that in turn restricts the applications of high-performance hard steels [1-5].

The new solution is the hot machining, which combines a heating source with conventional machining methods. Hot machining has emerged as a promising technique for hard materials that are difficult to cut. Besides, the workpiece is softened by heating during machining. It is showing a great promise in increasing the productivity and improving the quality due to a remarkable reduction in both cutting forces and tool wear [6-8].

Among hot machining processes is the laser assisted machining that combines laser beam technology with traditional machining methods such as turning and milling. Laser assisted machining is characterized by removal of controlled localized heated metal, which means maintaining the properties of the remaining bulk material. It is considered as a new and innovative way of machining the materials that are difficult to cut in the hardened state, such as ceramics, nickel based super alloys, titanium alloys and hardened ferrous alloys. Effectiveness of the laser assisted machining has already been demonstrated by different researchers [9-15]. Although scientific work on this area has contributed to understanding the process, there are still unresolved problems. Deep understanding of this process fundamentals including temperature distribution at material removal zone as a function of laser and machining parameters as well as material physical properties is of considerable importance regarding the machining quality.

It is extremely difficult or impossible to directly measure the parameters of interest such as heat depth or temperature distribution inside the workpiece during machining. In comparison with experimental work, modeling and simulation are of remarkable importance regarding accuracy and economy aspects. Numerical modeling has been proven to be an efficient method to understand physical phenomena and to predict control parameters of a certain process [16-18]. Mathematical models do not only allow an estimation of such parameters but also have the ability to lead into new directions due to the detailed analysis that can be achieved.

Mathematical modeling and simulation have been developed and used to understand and gain insight into the various mechanisms which take place during heating of a rotating workpiece using a moving laser beam. Transient three-dimensional model of a rotating workpiece undergoing laser heating and material removal was developed and used for different materials including mainly ceramics as well as titanium alloys [19-23]. However, research work for a moving Gaussian laser heat source to predict depth of heat-affected zone and temperature distribution in a Cartesian coordinates of hard steels is far from complete and more research work is required.

The current study aims at modeling and simulation of the laser assisted turning of hard steels for deep understanding of the temperature distribution at the material removal zone as a function of both laser and machining parameters.

2. MATHEMATICAL MODELING AND SIMULATION

2.1. Mathematical Modeling

The concept of laser-assisted machining of hard steels is based on using a laser as a heat source with the beam positioned directly in front of the cutting tool to heat and soften the work metal selectively before machining. In this case, the laser beam energy is absorbed by the workpiece surface and converted to thermal energy to produce rapid heating and softening of the part's upper layer reducing the cutting forces and thereby increasing the ease of material removal.

Schematic illustration of a rotated cylindrical workpiece subjected to laser heating is shown in **Figure 1**. To simulate the temperature distribution in this case, transient 3-dimensional numerical model was developed based on Rozzi et al model [21,22].

The laser heat source is assumed to have Gaussian distribution with a maximum heat flux at the laser beam spot center. The laser heat flux is a function of the laser power, the laser beam spot radius and the radial distance from the laser beam spot center, as shown in equation (1) [24].

$$q_l = \frac{2P}{\pi r_b^2} e^{-2r^2/r_b^2} , \qquad (1)$$

where q_l is the laser heat flux, P is the laser power, r_b is the radius of laser beam spot at the workpiece surface, and r is the radial distance from the laser beam spot center.

Assuming isotropic thermal conductivity and neglecting the convection effect of the laser gas jet, the heat is generated due to oxidation resulted from high temperature of the laser beam heating. The governing heat conduction equation in case of a rotated workpiece undergoing moving heat source is as following [21,22].

$$\frac{1}{r} + \frac{\partial}{\partial r}\left(kr\frac{\partial T}{\partial r}\right) + \frac{1}{r^2}\frac{\partial}{\partial \varphi}\left(k\frac{\partial T}{\partial \varphi}\right) + \frac{\partial}{\partial z}\left(k\frac{\partial T}{\partial z}\right) = \frac{\partial(\rho Cp\omega T)}{\partial \varphi} + \frac{\partial(\rho CpVzT)}{\partial Z} + \frac{\partial(\rho CpT)}{\partial t} \qquad (2)$$

To calculate the heat flux losses to ambient or surroundings, both convection and radiation boundary conditions are applied to the workpiece surface and the following equation was used.

$$q_s = h\left(T_{max} - T_{ambient}\right) + \varepsilon B\left(T_{max}^4 - T_{ambient}^4\right) \qquad (3)$$

The initial condition is T

$$T(\varphi, r, t = 0) = 300k \tag{4}$$

and the boundary conditions are as following:

- At the laser beam spot:

$$k\frac{\partial T}{\partial r}\text{ at surface} = \alpha_l q_l - q_{\text{conv}} - q_{\text{rad}} \tag{5}$$

- Off the laser beam spot:

$$k\frac{\partial T}{\partial r}\text{ at surface} = -q_{\text{conv}} - q_{\text{rad}} \tag{6}$$

where

$$q_{\text{conv}} = h_{\text{conv}}(T_{\text{max}} - T_{\text{ambient}}) \tag{7}$$

$$q_{\text{rad}} = \varepsilon B\left(T_{\text{max}}^4 - T_{\text{ambient}}^4\right) \tag{8}$$

T_{max} is the maximum surface temperature out of the laser beam spot.

T_{ambient} is the ambient temperature.

2.2. Simulation of the Model

For simulation of the above model, AISI51 50H hot worked tool steel was selected based on an available experimental data for this type of steel [20]. In this regard, a rotated workpiece with 32 mm diameter and 100 mm length subjected only to translating laser heating with no material removal was considered.

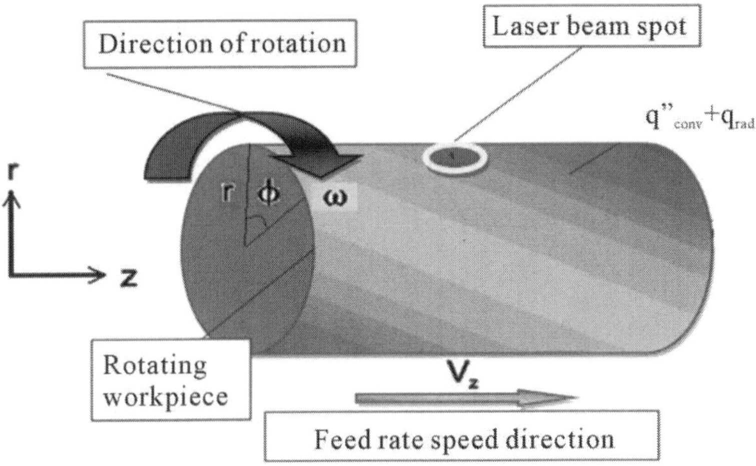

Figure 1. Schematic illustration of a rotated cylindrical workpiece subjected to laser heating.

Chuck holder of the workpiece has not been included in the model. As only the heat transfer process is analyzed, the workpiece divided into four regions including surface, interior and two sides. The summation of the cells is 288,234 cells including 878,799 faces and 302,596 nodes.**Figure 2** shows meshes created by Gambit-CFD for the used workpiece. Once the modeling and meshing have been done, the grid is imported into a finite volume software package (Fluent®), where three-dimensional single precision solver was used for analysis. The next step of the model development is the definition of the material properties as well as the definition of a moving Gaussian heat source. The thermo physical properties of AISI51 50H steel are given in **Table 1**. Since transient boundary conditions cannot be defined by the fluent user-interface, a User-Defined Function was created, using C language, (**Figure 3**) to enable applying a heat flux to a rotating cylindrical workpiece. This User-Defined Function is very important to customize the Fluent Code to calculate the heat flux as a particular modeling need. The heat flux is defined as a custom boundary profile that varies as a function of spatial co-ordinates and time. The established User-Defined Function is compiled at the run time by built-in complier/interpreter in the Fluent Code. Once the User-Defined Function is interpreted, the function is available in the graphical interface of the Fluent Code. This User-Defined Function is called at every time step, where the laser heat flux center is calculated as a function of time. The User-Defined Function loops cover all the cell faces on the surfaces, where the distance between the center of each cell face and the laser spot is calculated. If this distance is equal to or less than the laser spot radius then, the heat flux is calculated using equation (5). Otherwise Equation (6) is used in case of convection and radiation condition.

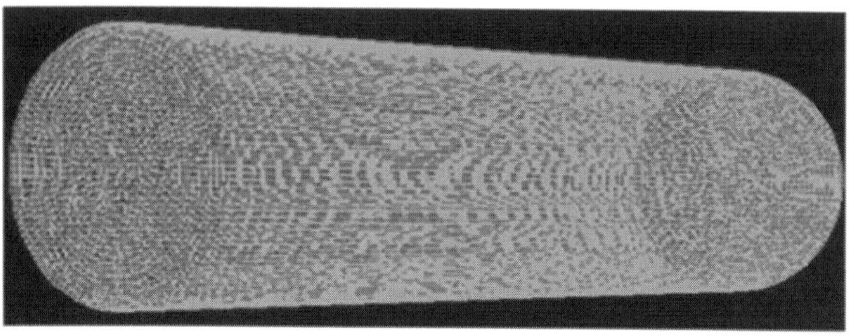

Figure 2. Meshes created by Gambit-CFD for the used workpiece.

2.3. Model Validation

Based on the above simulation, surface temperature of the used steel workpiece was calculated as a function of time at a specific distance behind the laser beam spot, which is corresponding to 30° angle from the laser beam. To ensure maximum uniform heat distribution without damaging/melting of the workpiece, the laser beam spot was defocused and kept constant at 5 mm diameter. Reflectivity of the material to the laser beam was considered as 50%.

Table 1. Thermo physical properties of AISI51 50H steel [20].

Property	Temperature	Value
Thermal conductivity, $k\left(\frac{w}{mK}\right)$	T = 300 k 300 < T(k) ≤ 1100 T > 1100 k	47.54 = 0.0275 * T(k) + 55.79 25.54
Specific heat, Cp (J/kg·K)	T = 300 k 300 < T(k) ≤1100 T > 1100 k	441.3 0.2195 * T(k) + 375.48 617
Density, $\rho\,(kg/m^3)$	Any temp.	7850
Reflectivity, α	Any temp.	0.5
Emissivity ε	Any temp.	0.9

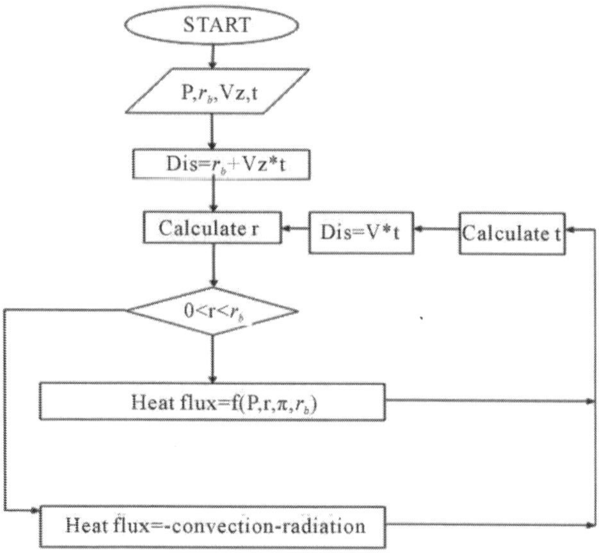

Figure 3. Flow chart of User-Defined Function created to enable applying a heat flux to a rotating cylindrical workpiece. P is the power input, r_b is the laser beam radius, Vz is the linear speed of laser beam, t is the interaction time, r is the distance from the laser beam spot center, d is the offset distance.

The results were compared with other published data of the same used steel, which included both experimentally measured and calculated values based on a transient, three-dimensional model [20]. This comparison disclosed a close agreement between both present and published results, as shown in **Figure 4**.

3. RESULTS AND DISCUSSION

The verified model was used for simulation of laser-assisted turning process of a different tool steel (AISI D2) having different dimensions. In this regard, a workpiece with 20 mm diameter and 300 mm length was considered. The thermo physical properties of this steel type are given in **Table 2**. The cutting depth was studied as a function of different laser-assisted turning parameters including preheating time, laser power, laser scanning speed and workpiece rotational speed. Then, the proper laser-assisted turning parameters were determined. In order to avoid damage of the laser head and for having sufficient access for handling both laser head and cutting tool, the cutting zone behind the laser head was set at 60° angle from the laser beam, as shown schematically in **Figure 5**.

It is speculated that the hardness of AISI D2 tool steel is decreased and subsequently its toughness is improved with heating above 540°C due to softening. The maximum cutting depth is decided based on this temperature for easy and successful cutting process. In other words, the maximum cutting depth is considered as the depth heated above 540°C. The temperature of the remaining bulk material is kept below 540°C to minimize its microstructure change and then to maintain its properties.

3.1. Effect of Preheating Time

For easy and successful cutting, it is of considerable importance to maintain uniform temperature distribution during the laser-assisted turning process. In this concern, the preheating time before starting metal removal has a remarkable effect. It is necessary to match the preheating time with the other parameters of the laser-assisted turning process so that steady-state temperature distribution is achieved, just at the beginning of metal removal. To ensure maximum uniform heat distribution, without damaging/melting the workpiece or tool, the laser beam spot was defocused and kept constant at 5 mm diameter. Reflectivity of the used steel to the laser beam was considered as 50%.

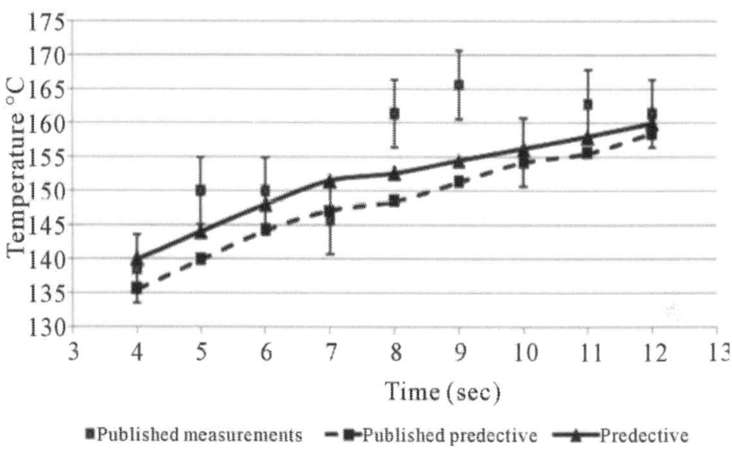

Figure 4. Comparison between the present results of calculated surface temperature as a function of time and other experimentally measured and calculated data [20]. Laser power: 500W; Laser scanning speed: 0.4 mm/s; Rotational speed: 6 rpm.

For clarifying the effect of preheating time, the surface temperature was calculated at different preheating times (10, 20, 30 sec) while other parameters were kept constant at 800 W laser power, 0.0 mm/sec laser scanning speed, 300 rpm rotating speed and 0.0 mm/sec feed rate. The laser power was selected based on an available CO_2 laser system having a maximum output power of 1000W. The rotational speed was selected based on previous accumulated practical machining experience. **Figure 6** shows an example of three-dimensional surface temperature distribution after 20sec preheating of 20 mm diameter workpiece with other parameters kept constant. It is found that 20 sec is the optimum preheating time for successful cutting since longer preheating time results in surface melting while shorter time results in insufficient heating for easy and successful cutting.

Table 2. Thermo physical property of AISI D2 steel [25].

°C / Property	20	200	400
Thermal conductivity. $k\left(\frac{w}{mk}\right)$	20	21	23
Specific heat, Cp (J/Kg·K)	460	460	460
Density, $\rho\left(\frac{kg}{m^3}\right)$	7600	7600	7600

3.2. Effect of Laser Power

To clarify the effect of laser power on the cutting depth, three-dimensional heat distribution was calculated as a function of laser power with other parameters kept constant at 20 sec preheating time, 0.8 mm/sec laser scanning speed, 300 rpm rotational speed and 0.8 mm/sec feed rate. It should be reported that 0.8 mm/sec feed rate was selected based on previous accumulated practical machining experience. Based on the selected feed rate, the laser scanning speed was set at the same value; 0.8 mm/sec.

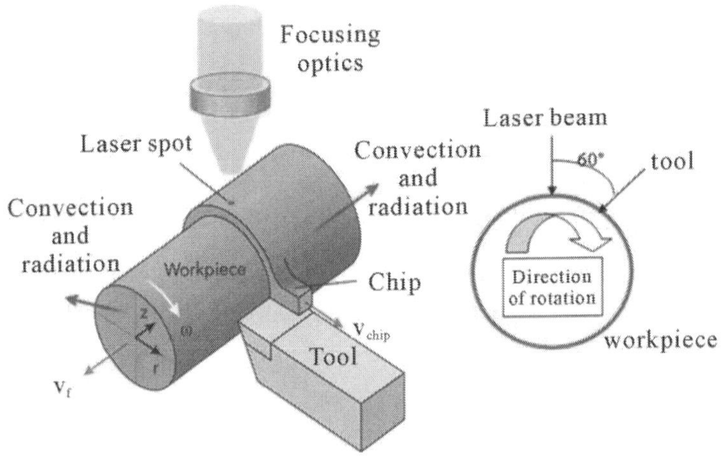

Figure 5. Schematic illustration of a laser-assisted turning process.

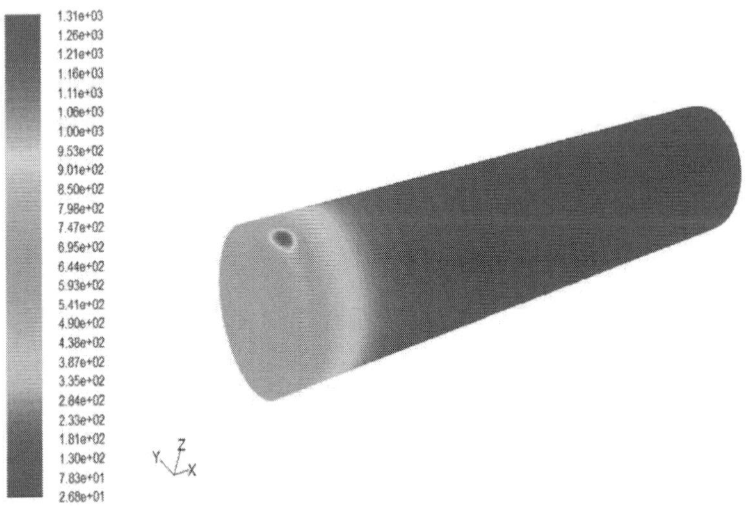

Figure 6. Three dimensional surface temperature distribution after 20 sec preheating of 20 mm diameter workpiece with other parameters kept constant at 800 W laser power, 0.0 mm/sec laser scanning speed, 300 rpm rotating speed and 0.0 mm/sec feed rate.

Figure 7 shows the temperature distribution through the depth as a function of the laser power with the other parameters kept constant. It can be noticed that laser power lower than 800 W has resulted in insufficient surface heating (450°C maximum). This means insufficient reduction in the hardness of the outer layer and then, difficult or unsuccessful cutting. Increasing the laser power to 800 W has resulted in increasing the surface temperature to 595.6°C, which is sufficient for easy/successful cutting. Based on the minimum temperature (540°C) required to reduce hardness for easy cutting, the maximum depth that can be successfully cut in this case is 1mm where proper temperature (546°C) is still existed. More increase in laser power (1000 W) has resulted in 735.63°C surface temperature and 540°C at 5 mm depth, which is still within the cutting range. In other words, increasing the laser power to 1000 W will increase the possible cutting depth to 5 mm. However, this cutting condition is not practically applicable since it is difficult to cut 5 mm depth in a one pass. For successful laser assisted turning process that means high cut quality with maintaining the properties of the remaining bulk material, the depth of the decreased hardness or softened layer should be completely removed in a one pass. It can be deduced that 800 W is considered as the optimum laser power regarding both quality and economy aspects.

3.3. Effect of Laser Scanning Speed

The effect of the laser scanning speed on the cutting depth was studied through calculation of three-dimensional heat distribution as a function of laser scanning speed with other parameters kept constant. In this concern, the laser scanning speed was varied between 0.2 and 1.0 mm/sec while the other parameters kept constant at 800 W laser power, 20sec preheating time, 300 rpm rotational speed and 0.8 mm/sec feed rate. **Figure 8** shows the temperature distribution through the depth as a function of the laser scanning speed with the other parameters kept constant. For laser scanning speed range of 0.2 - 0.6 mm/sec, the surface temperature of the workpiece is increased up to 1088°C due to massive or high heat input. The temperature is decreased up to 541°C at 10 mm depth, which indicates that the workpiece center is located within the easy cutting range. However, this cutting condition is not applicable due to difficulty in removing such thick layer (10 mm depth) in a one pass. In addition, massive heat input will result in deterioration of the properties of both the remaining bulk material as well as the cutting tool.

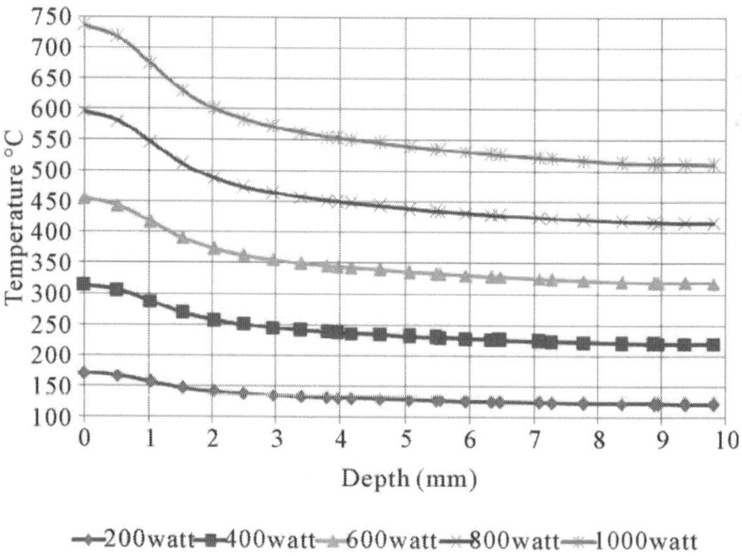

Figure 7. Temperature distribution through depth as a function of laser power with other parameters kept constant at 20 sec preheat time, 0.8 mm/sec laser scanning speed, 300 rpm rotational and 0.8 mm/sec feed rate.

Increasing the laser scanning speed to 0.8 mm/sec has resulted in acceptable heat distribution through the workpiece where surface temperature of 595.6°C was obtained. Based on the required minimum heating temperature (540°C) for easy cutting, the maximum depth that can be successfully cut is 1 mm. Increasing the laser scanning speed to 1 mm/s has resulted in a lower surface temperature (505°C), which means insufficient heating for successful cutting. It can be deduced that 0.8 mm/sec is considered as the optimum laser scanning speed regarding both quality and economy aspects.

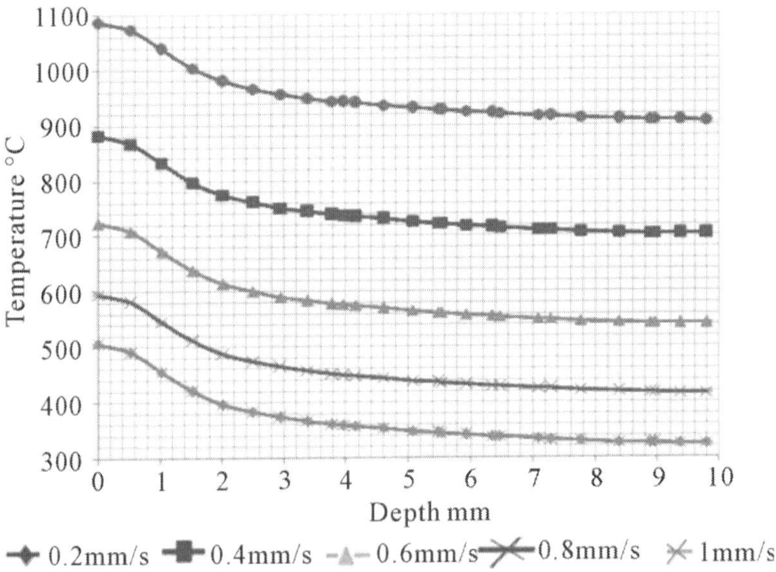

Figure 8. Temperature distribution through depth as a function of laser scanning speed with other parameters kept constant at 800 W laser power, 20 sec preheating time, 0.8 mm/sec feed rate and 300 rpm rotational speed.

3.4. Effect of Workpiece Rotational Speed

Three-dimensional heat distribution was calculated for different rotational speeds with the other previously obtained optimum parameters kept constant. In this concern, the rotational speed was varied between 100 and 900 rpm while the other parameters kept constant at 800 W laser power, 20 sec preheating time, 0.8 mm/sec laser scanning speed and 0.8 mm/sec feed rate. **Figure 9** shows the temperature distribution through depth as a function of the rotational speed with other parameters kept constant. For 100 rpm rotational speed and based on minimum heating temperature (540°C) required for successful cutting, a depth of 1.5 mm should be completely cut or removed in a one pass in order to avoid deterioration of the properties of the remaining bulk material. Much better heat distribution through the depth was achieved with 300 rpm rotational speed, where 1 mm depth can be easily or successfully cut without deteriorating the properties of the remaining bulk material.

Increasing the rotational speed to 500 and 700 rpm has resulted in lower surface temperatures (570°C - 554.57°C), which are still sufficient for cutting. However, the maximum depth that can be easily or successfully cut is limited to 0.5 mm based on the required minimum heating temperature (540°C). This relatively shallow cut depth is not economically reasonable. More increase in the rotational speed to 900 rpm has resulted in insufficient heating for easy/successful cutting of even 0.5 mm depth. This is due to insufficient heating temperature (534.7°C) at this depth.

Based on the above all results, the most suitable parameters for laser-assisted turning of the concerned steel are 800 W laser power, 5 mm laser beam spot diameter, 20 sec preheating time, 0.8 mm/sec laser scanning speed, 300 rpm rotational speed and 0.8 mm/sec feed rate. These parameters ensure successful cutting of 1 mm depth without deteriorating the properties of the remaining bulk material.

4. CONCLUSIONS

A transient, three-dimensional model was developed to predict the temperature distribution of a rotated cylindrical steel workpiece subjected to a localized heating using a moving Gaussian laser beam. In this concern, a UserDefined Function (C program) was established to overcome the problem of the definition of a moving Gaussian heat source. This User-Defined Function was compiled into a finite volume software package (Fluent®), where three-dimensional single precision solver was used for analysis. Based on this model, simulation of the surface temperature of 32 mm diameter workpiece of AISI51 50H steel was performed as a function of time at a specific distance behind the laser beam spot, which is corresponding to 30° angle from the laser beam. The simulation results were compared with other published data of the same steel type where a close agreement was obtained.

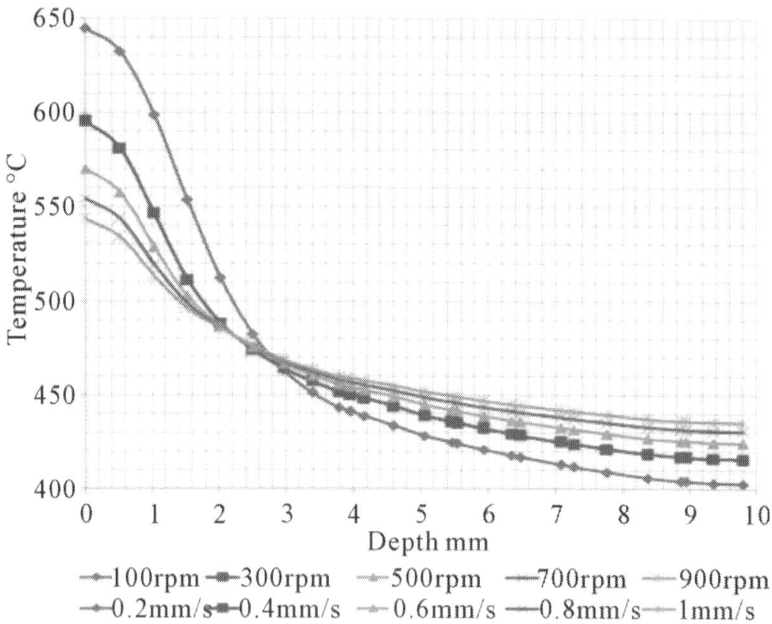

Figure 9. Temperature distribution through depth as a function of rotational speed with other parameters kept constant at 800 W laser power, 20 sec preheating time, 0.8 mm/sec laser scanning speed and 0.8 mm/sec feed rate.

The verified model was used for simulation of laser assisted turning of 20 mm diameter workpiece of AISI D2 tool steel. The cutting zone behind the laser beam was set at a distance corresponding to 60° angle from the laser beam for having sufficient access for handling both laser head and cutting tool. Prediction of the cutting depth was done as a function of different lasers and machining parameters. The results indicated that the optimum parameters for successful laser-assisted turning process of the concerned steels are 800 W laser power, 5 mm laser beam spot diameter, 20 sec preheating time, 0.8 mm/sec laser scanning speed, 300 rpm rotational speed and 0.8 mm/sec feed rate. These parameters ensure easy /successful cutting of 1 mm depth in one pass without deteriorating the properties of the remaining bulk material.

It can be deduced that the developed model might provide a useful tool for online process control of different steel types regardless of their physical properties and geometry.

ACKNOWLEDGEMENTS

The authors would like to thank Prof. G. Baumann, Head of Mathematics Department at Germany University in Cairo for his valuable advice and fruitful discussion on mathematical modeling.

REFERENCES

1. R. Poggie, "Wear of Materials," Journal of Materials Science and Technology, Vol. 7, No. 2, 1991, pp. 387-406.

2. Lima, "Hard Turning: AISI 4340 High Strength Low Alloy Steel and AISI D2 Cold Work Tool Steel," Journal of Materials Processing Technology, Vol. 169, No. 3, 2005, pp. 388-395.

3. S. Herbert, "Metal Cutting and High Speed Machining," 2nd Edition, Springer, New York, 1994.

4. G. Brain, "Manufacturing Surface Technology," 3rd Edition, Taylor and Francis Inc., London, 2001.

5. L. Bourithisa, "Comparison of Wear Properties of Tool Steels AISI D2 and O1 with the Same Hardness," Tribology International, Vol. 39, No. 6, 2006, pp. 479-489.

6. L. Zler, "Theoretical and Experimental Determination of Tool Life in Hot Machining of Austenitic Manganese Steel," International Journal of Machine Tools and Manufacture, Vol. 41, No. 2, 2001, pp. 163-172.

7. J. A. Sanchez, "Plasma Assisted Milling Heat-Resistant Superalloys," Journal of Manufacturing Science and Engineering, Vol. 126, No. 2, 2004, pp. 274-285.

8. S. Sun, M. Brandt and M. Dargusch, "Thermally Enhanced Machining of Hard-to-Machine Materials," International Journal of Machine Tools and Manufacture, Vol. 50, No. 8, 2010, pp. 663-680.

9. M. Anderson, R. P. Yung and C. Shin, "Laser-Assisted Machining of Inconel 718 with an Economic Analysis," International Journal of Machine Tools and Manufacture, Vol. 46, No. 14, 2006, pp. 1879-1891.

10. S. Rajagopal, "Machining Aerospace Alloys with the Aid of a 15kW Laser," Journal of Applied Metal Work, Vol. 2, No. 3, 1982, pp. 170-184.

11. S. Sun, "Parametric Investigation of Laser-Assisted Machining of Commercially Pure Titanium," Advanced Engineering Materials, Vol. 10, No. 6, 2008, pp. 565-572.

12. B. Yang, "Laser-Assisted Milling of Silicon Nitride Ceramic :A Machinability Study," International Journal of Mechatronics and Manufacturing Systems, Vol. 1, No. 1, 2008, pp. 116-130.

13. F. E. Pfefferkorn, et al., "Laser Assisted Machining of Magnesia-Partially Stabilized Zirconia," Journal of Manufacturing Science and Engineering, Vol. 126, No. 1, 2005, pp. 346-357.

14. P. Dumitrescu, et al., "High-Power Diode Laser Assisted Hard Turning of AISI D2 Tool Steel," International Journal of Machine Tools and Manufacture, Vol. 46, 2006, pp. 2009-2016.

15. N. Samant and N. B. Dahotre, "Laser Machining of Structural Ceramics—A Review," Journal of European Ceramic Society, Vol. 29, No. 6, 2009, pp. 969-993.

16. J. Mazumder and W. M. Steen, "Heat Transfer Model for CW Laser Materials Processing," Journal of Applied Physics, Vol. 51, No. 2, 1980, pp. 941-947.

17. K. S. Yeung and P. H. Thornton, "Transient Thermal Analysis of Spot Welding Electrodes," Welding Journal, Vol. 78, No. 1, 1999, pp. 1S-8S.

18. D. Mandelprot, "Modeling the Fluid-Flow in Laser-Beam," Welding Journal, Vol. 65, No. 3, 1986, pp. 167S-174S.

19. X. W. Shen and L. Shuting, "Thermal Modeling and Experimental Investigation for Laser Assisted Milling of Silicon Nitride Ceramics," Journal of Manufacturing Science and Engineering, Vol. 131, No. 5, 2009, pp. 1254- 1274.

20. R. Patwa and Y. C. Shin, "Predictive Modeling of Laser Hardening of AISI51 50H Steels," International Journal of Machine Tools and Manufacture, Vol. 47, No. 2, 2007, pp. 307-320.

21. J. C. Rozzi, et al., "Transient, Three-Dimensional Heat Transfer Model for Laser Assisted Machining of Silicon Nitride: I.

Comparison of Predictions with Measured Surface Temperature Histories," International Journal of Heat and Mass Transfer, Vol. 43, No. 8, 2000, pp. 1409- 1424.

22. J. C. Rozzi, F. P. Incropera and Y. C. Shina, "Transient, Three-Dimensional Heat Transfer Model for the Laser Assisted Machining of Silicon Nitride: II. Assessment of Parametric Effects," International Journal of Heat and Mass Transfer, Vol. 43, No. 8, 2000, pp. 1425-1437.

23. P. A. Rebro, Y. C. Shin and F. P. Incropera, "Design of Operating Conditions for Crack Free Laser-Assisted Machining of Mullite," International Journal of Machine Tools and Manufacture, Vol. 44, No. 7-8, 2004, pp. 677- 694.

24. W. A. Chang, "A Study on Heat Source Equations for the Prediction of Weld Shape and Thermal Deformation in Laser Micro Welding," Metallurgical and Materials Transactions B-Process Metallurgy and Materials Processing Science, Vol. 33, No. 5, 1991, pp. 757-764.

25. M. H. Mohamed, "Laser Assisted Machining of Tool Steel," M.Sc. Thesis, Military Technical College, Cairo, 2007.

6

FLOW-OPTIMIZATION TO ENHANCE GAS QUENCHING EFFICIENCY FOR HELICAL GEARS SPECIMEN

Thibaud Bucquet, Udo Fritsching

Process & Methods Department, IWT Bremen, Bremen, Germany

ABSTRACT

Numerical simulations based on a conjugate heat transfer solver have been carried out to analyze various gas quenching configurations involving a helical gear streamed by an air flow at atmospheric pressure in a gas quenching chamber. In order to optimize the heat transfer coefficient distribution at key positions on the specimen, configurations involving layers of gears and flow ducts comprising single to multiple gears have been simulated and compared to standard batch configurations in gas quenching. Measurements have been performed covering the local heat transfer for single gears and batch of gears. The homogeneity of the heat transfer coefficient is improved when setting up a minimal distance between the gears (batch density) and when introducing flow ducts increasing the blocking grade around the gears. An offset between layers of the batch as well as flow channels around the gears plays a significant role in increasing the intensity and the homogeneity of the heat transfer in gas quenching process.

KEYWORDS

Heat Treatment, Gas Quenching, Helical Gears, Computational Fluid Simulation, Flow Optimization

1. INTRODUCTION

In the automotive industry, in order to achieve a sufficient hardness distribution (corresponding to a martensitic microstructure) or shape quality [1], specific heat treatment processes are applied in the manufacturing process of parts. Gas quenching has been developed over the past years as an environmentally and economically friendly solution to replace conventional liquid or salt-based quenching [2]. Whereas conventional oil-based quenching provides heat transfer coefficients ranging from 1000 to 5000 $W/(m^2 \cdot K)$ [3], the polluting effect of oil requires a specific safety level and cleaning the specimen after treatment [4]. Fluid-based quenching has the major inconvenience of facing changes in its liquid phase-state and heat transfer; from film boiling to bubble boiling and finally pure convection [5]. Gas quenching defined in [6] - [8] aims at reducing the distortion potential caused by the high variations in the heat transfer intensity due to the liquid evaporation occurring during conventional liquid-based quenching.

In comparison to quenching techniques using liquids, the impact of the gas flow on the heat transfer between the quenching gas and the element to be quenched is of major importance [9]. This influence can be scaled according to the dimension of the system: at the micro scale, the solid workpiece is streamed by a complex gas flow depending on the structure of the batch (meso scale) comprising many layers of workpieces as well as racks creating a pressure drop to the incoming flow. The use of flow control elements at meso scale has been discussed in [10] for application to shaft-shaped specimen. At the macro scale of the quenching chamber, particular flow patterns arise, depending on the technical aggregates (heat exchangers, recirculation of the gas) and the geometry of the chamber (dead flow zones, chamber inlet/outlet).

Heat transfer in industrial gas quenching is described by the heat transfer coefficient h, independent of the quenching medium temperature and the specimen temperature and material properties. In industrial gas quenching, the heat transfer coefficient has been derived as [11] :

$$h = c_2 \cdot \mu_\infty^{0.7} \cdot p^{0.7} \cdot l^{-0.3} \cdot \eta^{-0.39} \cdot c_p^{-0.31} \cdot \lambda_{Gas}^{0.69}, \qquad (1)$$

where c_2 is a factor related to the quenching chamber geometry.

1.1. Measurement in Industrial Gas Quenching

The heat transfer in gas quenching is analyzed using a local and an integral method to determine the heat transfer coefficient on helical gear specimen. The local determination of the heat transfer coefficient is done through measurement of the local heat flux on the specimen surface using a 5 × 5 mm Captec sensor with integrated thermocouple to monitor the specimen surface temperature [12]. The integral determination of the heat transfer is performed assuming the lumped capacitance approach described below for the gear to be investigated. The lumped capacitance assumption requires a Biot-number

$$B_i = \frac{hl}{\lambda_s} < 0.1, \qquad (2)$$

where l is taken as the streamwise length of the gear [13].

Quenching conditions involving small heat transfer coefficients for material of high heat conductivity (copper, aluminum) lead the effect of internal conduction to overcome the effect on the external heat transfer by convection so that the temperature distribution is considered uniform at any time in the solid. Thus the integration of the heat equilibrium equation,

$$hA\left(T_w - T_\infty\right) = -\rho C_p \frac{\partial T}{\partial t} V \tag{3}$$

can be reduced to a time-depending integration.

In that case, the temperature can be monitored at any point within the specimen and the integral or mean heat transfer coefficient may be calculated. Three specimen helical gears of thicknesses 20, 25 and 30 mm made of EN-AW6082 to minimize the Biot-number have been investigated during the quenching process.

The gas velocity in gas quenching process is measured using a pressure-based 7-hole, pitot probe whose orientation and calibration allow the measurement of the 3 velocity components in a 72° cone in relation to the incoming flow [14].

1.2. Simulation in Industrial Gas Quenching

The simulation of the heat transfer process in gas quenching requires accurate modelling of the quenching environment (flow velocity, turbulence and chamber geometry), at the macro and meso scales. The accurate modelling of the micro scale, that is, the helical gear itself, ensures a valid computation of the heat transfer coefficient on its surface. The simulation library Open FOAM 2.0.1 provides an existing steady-state, coupled heat transfer simulation solver. The computation of the gas flow is based on Reynolds-averaged Navier-Stokes (RANS) equations using the SIMPLE-algorithm (Semi-IMplicit Pressure-Linked Equations) and the energy equation for the fluid. It provides the temperature field around the solid specimen that is used to solve the heat conduction pro- cess for the selected regions to be quenched.

Heat transfer problems simulations at large scale are ([15] [16]) a challenging task when defining the grid accuracy around the specimen in relation to a relevant turbulent model. [17] provided an improvement of the k-ω SST turbulence model using near-wall functions increasing the independency of the turbulence model to the first near-wall grid point distance y+. In addition to combining the k-ω turbulence model for y+ < 200 with the free-stream k-ε model for y+ > 200, near-wall functions provide a blending of the viscous and log formulations of the flow depending on the first near-wall grid point. For applications in industrial heat transfer process, this near-wall treatment has demonstrated its effectiveness for grid structures at 100 > y+ > 0.2 [18] [19].

While the gas quenching chamber geometry is directly meshed within Open FOAM, flow ducts and the helical gears are implemented as. stl geometries in the mesh with finer resolution, so that the average grid cell number remains below 30 million cells. Inlet boundary conditions are fitted to the complex flow from the heat exchanger as velocity and turbulence components that are extrapolated from measurements in order to accurately model real gas quenching conditions. Figure 1 illustrates simulation results from steady-state coupled heat transfer with respect to measurements at 3 different positions (front side, tooth and rear side) on the helical gear using the heat flux sensors. A grid independency test has been carried out to validate the choice of the turbulence model. The various grid refinement levels are indicated in Figure 1 (right) from finer (green) to coarser (red) meshes. As y+ values from the viscous to the log-layer are used, the heat transfer coefficient remains almost constant with similar values as in the measurements. Local variations of the heat transfer coefficient are observed in areas undergoing a strong pressure gradient (recirculating flow after impinging onto the top of the gear).

2. INVESTIGATION

2.1. Flow Characteristics in Top-to-Bottom Flow Quenching Chambers

Top-to-bottom flow quenching chambers are found in industry for quenching process of gears [21]. These configurations ensure a homogeneous distribution of the main flow on the first layer of the batch, impinging directly onto the top side of the gears. The application of top-to-bottom gas quenching is done for single parts [22], 2- dimensional [21] layers of parts and conventional 3-dimensional [7] batches. The structure of a typical top-to- bottom quenching chamber is presented in Figure 2, with the investigated quenching chamber displayed on the left and a simplified sectional view of a top-to-bottom quenching chamber on the right.

Flow distribution characteristics in the gas quenching chamber are evaluated using the pressure probe. The probe tip is pointed towards the main flow direction (along the z-axis). The probe is mounted on a positioning system scanning the investigated planes (1 - 3) shown in Figure 3. In an empty quenching chamber (without load), the flow characteristics have been evaluated using air flow at atmospheric pressure. Three planes are measured, Plane 1 (Figure 3) corresponds to the inlet flow condition, Plane 2 is the place above the batch and Plane 3 is the flow condition when reaching the outlet.

The comparison of the velocity distributions is presented in Figure 4, for planes 1 to 3 respectively from the top (inlet/heat exchanger) to the bottom (outlet) of the chamber. While the gas velocity distribution on Plane 1 obtains locally inhomogeneous distribution, presenting horizontal patterns where the velocity magnitude varies from high to low values (attributed to the heat exchanger located above this plan), the homogeneity of the distribution increases as the gas flows downwards. On Plane 3 (outlet), higher velocities are found in the front and the rear area of the chamber, where respectively the quenching chamber and furnace doors are located.

These characteristics of top-to-bottom quenching chambers are due to the void or dead zones that are located besides the plane.

Numerical simulations have been performed based on the gas velocity distribution on Plane 1 as boundary conditions. The main velocity component (along the z-axis), the coefficient of variation, defined as VarCo = Sig/X and the turbulent kinetic energy have been detected and compared to simulation results for the empty flow chamber. A simulation involving a two-layer batch of helical gears has been related to the simulation/measure- ment in the empty chamber. The results are shown in Figure 5. The blue lines in all three diagrams display the empty chamber and the red lines the chamber with a batch. Empty markers indicate measurement and simulation in the empty chamber while half-full ones indicate simulations in batch-mode. Considering the streamwise vetains a batch. Supposing the mass flow rate to be constant through the chamber, this decreasing value is due to the diversion of the flow on top of the batch, thus transmitting the flow momentum in the transverse directions. The increasing value in the center as no flow resistance is encountered is also described in [23].

The increase in streamwise velocity for both empty cases in an empty chamber is due to the increase in flow uniformity shown in Figure 4, as the influence of the transverse velocity components reduces from the heat exchanger to the chamber outlet. This trend is also seen within the coefficient of variation (VarCo), increasing in the simulated case with batch. The complex batch structure is influencing the flow by increasing the velocity fluctuations while the uniformization of the flow seen in the empty chamber is characterized by a decreasing coefficient of variation. The turbulent kinetic energy (tkE) demonstrates an almost constant behavior, though the flow is becoming more homogeneous. Considering the chamber with batch, the turbulent kinetic energy is almost constant above the batch and strongly decreases through the batch itself, as the large turbulent eddies size decreases through the batch complex structure.

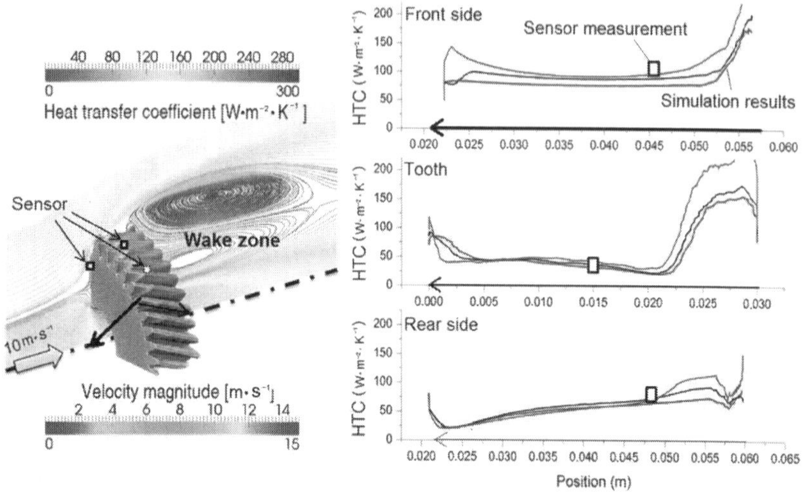

Figure 1. Flow around a helical gear; velocity magnitude and heat transfer coefficient distribution on the surface from simulation (left) and heat transfer coefficient distribution over the tooth length from simulation and measurement (right).

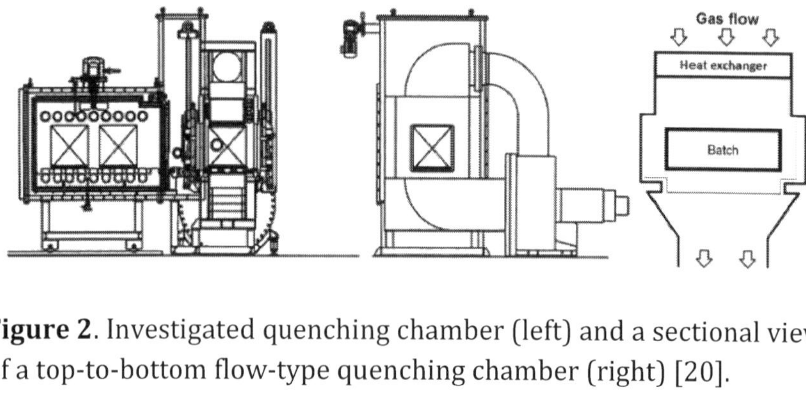

Figure 2. Investigated quenching chamber (left) and a sectional view of a top-to-bottom flow-type quenching chamber (right) [20].

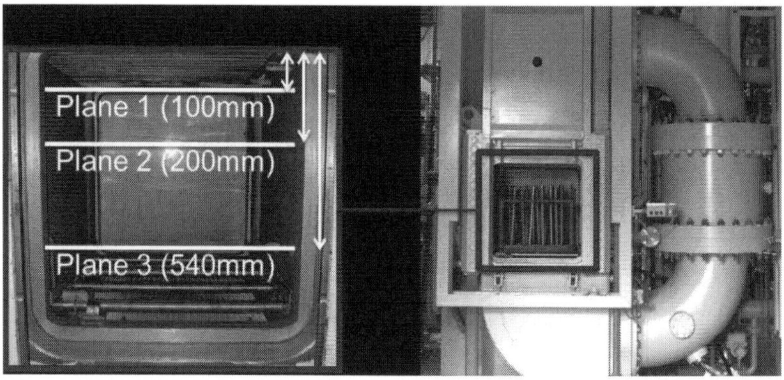

Figure 3. Position of the planes in the empty quenching chamber (left) compared to a batch located in the quenching chamber (right).

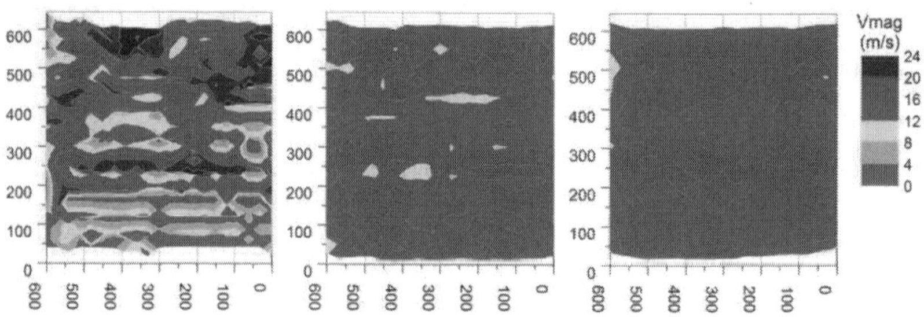

Figure 4. From left to right: velocity magnitude distribution in sections of the quenching chamber from top to bottom; measured.

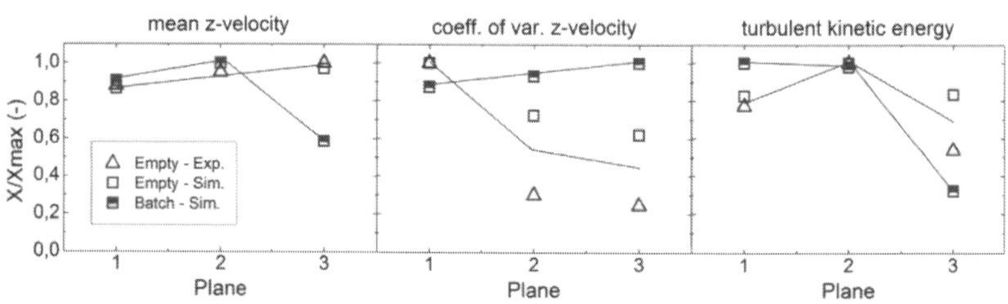

Figure 5. Results of the z-velocity, coefficient of variation and turbulent kinetic energy at various planes (1 to 3, top to bottom); measured and simulated.

2.2. Influence of Flow Ducts on the Heat Transfer

Simulations have been performed for a single helical gear in various flow configurations involving flow ducts around the specimen. The influence of the flow duct dimension on the local heat transfer at 4 key positions of the helical gear is investigated. The top surface (disc front) and the bottom surface of the gear (disc gear) as well as the streamed surface (tooth front) and the reverse side of the tooth (tooth rear) are investigated as the intensity and the uniformity of the heat transfer coefficient distribution over those surfaces play a major role on achieved quality of the workpiece in the heat treatment process. Figure 6 presents the results of the simulation involving configurations of flow ducts as: 1) without flow duct; 2) with a 160 mm-diameter cylindrical flow duct, then varying the diameter to 3) 140 mm and 4) 130 mm.

The velocity distribution around the flow duct and the gear is displayed around the gear. Such a recirculation area (Figure 1) behind the helical gear without flow duct is also observed in [15]. Introducing a flow duct and reducing its diameter decreases the size and the intensity of the recirculation zone (Figure 6). The velocity is increasing and more uniformly distributed in the tooth area. In Figure 7, the effect of flow uniformization and intensification correlates positively with an improved heat transfer.

The heat transfer on the specimen surface is evaluated in Figure 7 by the mean heat transfer coefficient left and the coefficient of variation right. The introduction of flow ducts intensifies the heat transfer except for the rear surface of the gear, with the smallest cylinder diameter obtaining the best improvement. The homogeneity of the heat transfer distribution (low coefficient of variation) is found best for the 130 mm-diameter cylinder. In the flow ducts investigation at batch level, the 140 mm-diameter cylinder was used as compromise between high and homogeneous heat transfer coefficient distribution for the investigated regions of the gear.

2.3. Batch Density

Considering a single layer of helical gears in gas quenching, the packing density of the batch has been evaluated, aiming at obtaining a homogeneous local heat transfer coefficient distribution at all gear wheels. At a high batch density, the distance between a gear and its neighboring specimen is low, as represented in Figure 8, so that the local flow-interaction between gears is high, leading to stronger variations of the heat transfer coefficient from one tooth to another as the density increases. On the other hand, a too large distance between gears thus reducing the batch density is energy inefficient.

Therefore, the influence of the distance between gears is investigated for the two sides of the critical gear tooth situated closest to a neighboring gear (Figure 8). The gear side exposed to the flow is represented in red while the rear side is in blue. For gear distances ranging from 0 (gears in contact) to 20 mm, the heat transfer coeffi cients are plotted in Figure 8 (right) from top to bottom. At the two gear sides (front and rear), a good homogeneity and high quenching intensity are found for small distances (0 and 2 mm), while at increased distances, larger heat transfer coefficients at the top and eventually bottom part of the tooth are formed.

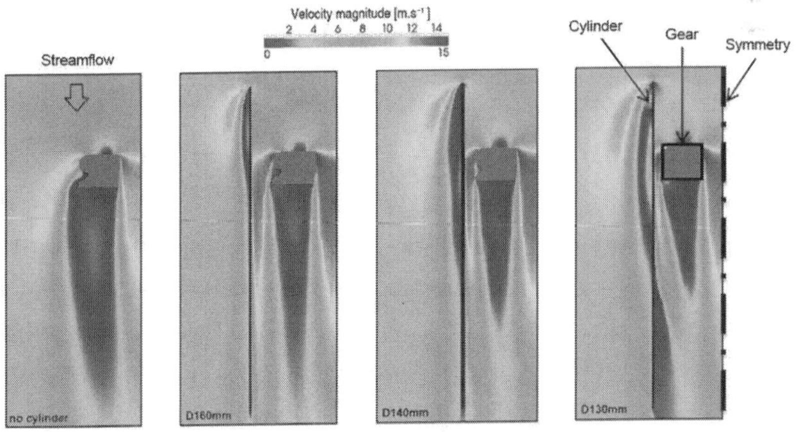

Figure 6. Velocity magnitude distribution around a helical gear with flow-duct configurations of reducing diameters; simulated.

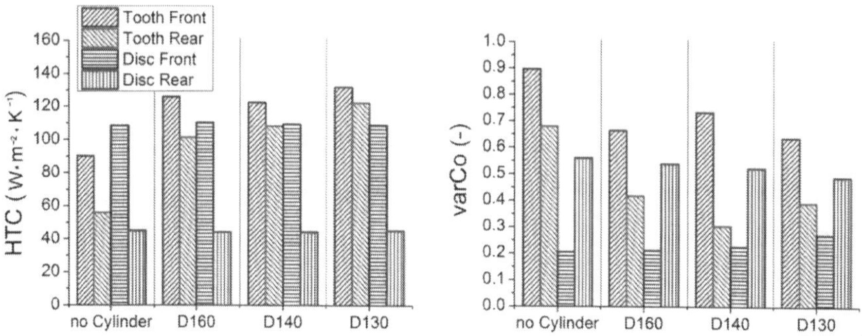

Figure 7. Heat transfer coefficient (left: average, right: coefficient of variation) at different locations for the 4 flow-duct configurations; simulated.

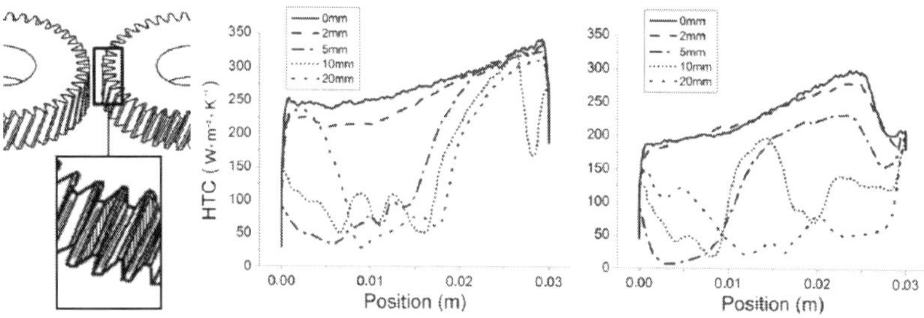

Figure 8. Local heat transfer coefficients along the front (left) and rear (right) side of a helical gear tooth for various batch densities (distance between gears in mm); simulated.

The flow is producing intense recirculations at higher distances, thus creating intensive heat exchange at the top, low heat exchange into the eddy in the middle of the tooth and eventually reattaching at the bottom, thus re-intensifying the heat transfer. The influence of the gear distance in the batch is displayed in Figure 9 (left) with the averaged heat transfer coefficient and corresponding standard deviations. The influence of the distance between the gears is large in the tooth area while the top and bottom sides of the gear are undergoing a less severe variation of the heat transfer intensity. As also confirmed by simulation in Figure 9 (right), a minimal gear distance between 5 and 10 mm is requested in order to ensure the most homogeneous quenching in all regions of the helical gear.

2.4. Flow Ducts Integration in Batch Configurations

Three configurations of batches involving staggered helical gears have been investigated using flow and heat transfer simulation as seen in Figure 10 (left). Confining flow ducts around helical gears may improve the quenching homogeneity in key regions of the workpiece such as the teeth. A 200 mm-height cylinder as shown in Figure 6 was investigated comprising 3 helical gears per cylinder unit. As seen in Figure 10, the integral heat transfer coefficient severely drops from the first layer to the second and third layer due to the blocking caused by the first layer in the confinement. In comparison to 200 mm-height, the 30 mm-height confinement allows a partial regeneration of the impacting flow from layer to layer. The highest heat transfer coefficients are found on the first layer with a progressive decrease to the lower layers.

An offset between parts is used (transverse in one direction) on the second layer for the third configuration of the batch in Figure 10 (bottom) utilizing the dynamic of the main flow between gears in subsequent layers. Combined with the 30 mm-height confinement, this configuration demonstrates the best results in terms of intensity and homogeneity of quenching. When considering the local quenching of the front side of a tooth from the first layer in Figure 11, the best compromise between high quenching intensity and high local homogeneity is achieved for quenching configurations with 30 mm-height cylindrical confinement.

The influence of confinement on the specimen heat transfer in gas quenching involving 2 to 3 layers in a batch with second-layer offset is summarized in Figure 12. The 30 mm-height flow ducts provides best results for the three layers in terms of quenching homogeneity and intensity. The third layer demonstrates good homogeneity but fails to reach the same level of quenching intensity when introducing a 1-directional offset at the second layer.

3. CONCLUSION

Measurements and simulations for evaluation of quenching heat transfer in batch configurations in industrial gas quenching of helical gears have been reported. A combination of adapted flow confinements with a layer disposition involving an offset of the second layer is improving quenching homogeneity and intensity of the quenching for the whole batch. The introduction of confinements improves the local homogeneity of the heat transfer coefficient in the tooth region of the helical gear specimen.

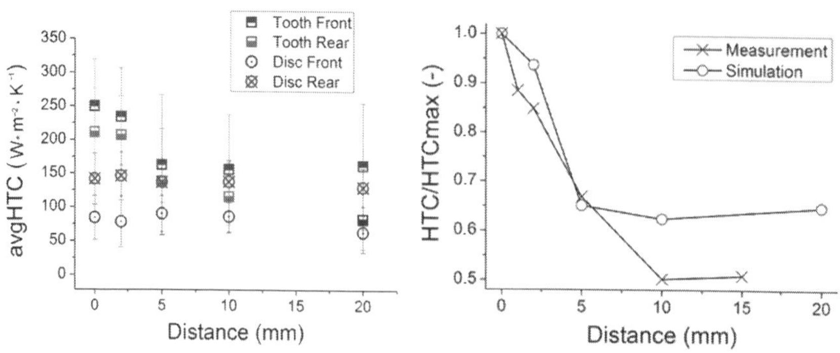

Figure 9. Simulated average heat transfer coefficients of a helical gear (left) and comparison (right) between measurement and simulation as a function of the batch density for the averaged heat transfer coefficient over the closest tooth region.

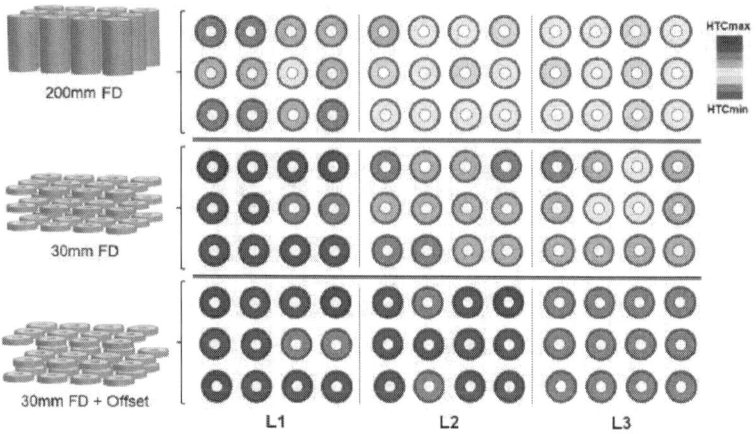

Figure 10. Average heat transfer coefficients at each layer (from left to right) for various batch configurations (from top to bottom); simulated.

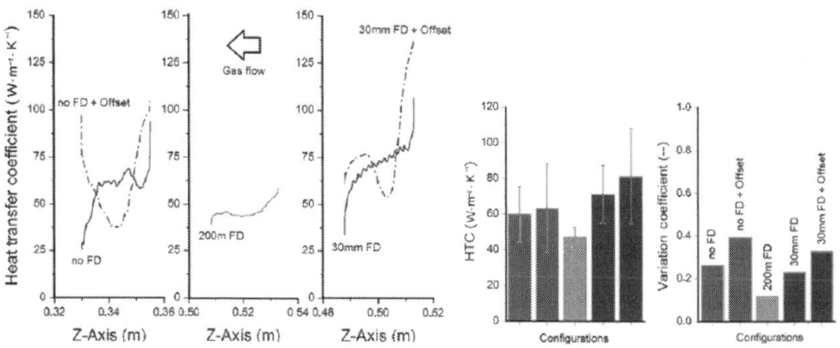

Figure 11. Local heat transfer coefficient along a gear tooth (left) with average and coefficient of variation (right) for various batch configurations; simulated.

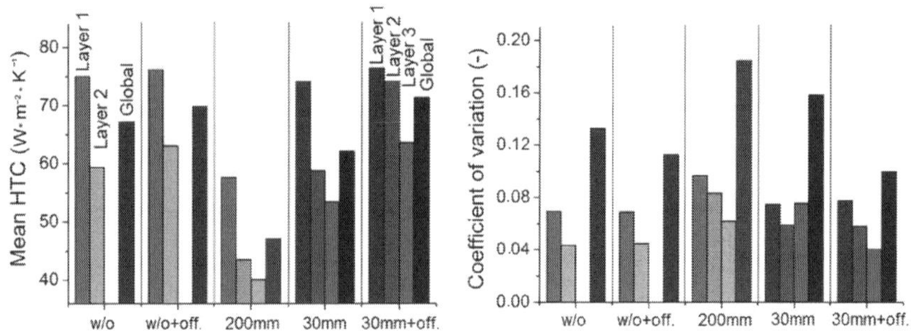

Figure 12. Mean and coefficient of variation for the integral heat transfer coefficients of the 2 and 3 layer batch configurations (without flow duct, 2nd layer offset without flow duct, with 200 mm-height flow duct, with 30 mm-height flow duct, 2nd layer offset with 30 mm-height flow duct); simulated.

ACKNOWLEDGEMENTS

The authors gratefully acknowledge support from the Forschungsgemeinschaft Industrieofenbau (FOGI e.V.) and AichelinGes.mbH, Mödling (Austria) for offering the possibility to perform in-situ test in industrial furnaces.

Thibaud Bucquet, Udo Fritsching (Arbeitsgemeinschaft Industrieller for Schungsvereinigungen "Otto von Guericke" e.V.) through financial resources from the BMWi (Bundesministerium für Wirtschaft und Technologie).

REFERENCES

1. Clausen, B., Frerichs, F., Kohlhoff, T., Lübben, T., Prinz, C., Rentsch, R., Sölter, J., Surm, H., Stöbener, D. and Klein, D. (2009) Identification of Process Parameters Affecting Distortion of Disks

for Gear Manufacture—Part II: Heating, Carburizing, Quenching. Materialwissenschaft und Werkstofftechnik, 40, 361-367.

2. Liscic, B., et al. (2011) Quenching Theory and Technology. Taylor & Francis, London.

3. Totten, G.E., Bates, C.E. and Clinton, N.A. (1993) Handbook of Quenchants and Quenching Technology. ASM Inter- national, Geauga County.

4. Altena, H., Schobesberger, P. and Schrank, F. (2006) Moderne Gas-Aufkohlungstechnik für die Automobilindustrie. Gas, Wärme International, 55, 484-489.

5. Stark, P. and Fritsching, U. (2011) Modeling and Simulation of Film and Transitional Boiling Processes on a Metallic Cylinder during Quenching. Journal of ASTM International, 8, 61-80.

6. Stolar, P., Altena, H., Jurči, P., Klima, F. and Honzik, O. (2001) Distortion of Gear Wheels After Quenching in Gas and Oil. Proceedings of the 8th Seminar of the International Federation for Heat Treatment and Surface Engineering, IFHTSE 2001, Dubrovnik-Cavtat, 12-14 September 2001, 95-102.

7. Altena, H., Jurci, P. and Stola, P. (2004) Gas and Oil Quenching Effects on Gear Distortion. Industrial Heating, 71, 45- 48.

8. Atraszkiewicz, R., Januszewicz, B., Kaczmarek, Ł., Stachurski, W., Dybowski, K. and Rzepkowski, A. (2012) High Pressure Gas Quenching: Distortion Analysis in Gears after Heat Treatment. Materials Science and Engineering: A, 558, 550-557.

9. Schüttenberg, S., Frerichs, F., Hunkel, M., Fritsching, U. and Mayr, P. (2004) Verzugskompensation Mittels Gasabschreckung in Flexiblen Düsenfeldern. HTM Härtereitechnische Mitteilungen, 59, 185-192.

10. Schmidt, R.-R. (2013) Zur Thermo-Fluid-Dynamik beim Hochdruckgasabschrecken: Experimentelle und Numerische Analyse der Hochdruckgasabschreckung Metallischer Bauteile zur Steigerung von Prozesshomogenität und-Intensität. Doctoraldissertation, Universität Bremen, Bremen.

11. Löser, K., Heuer, V. and Schmitt, G. (2005) Auswahl geeigneter Abschreckparameter für die Gasabschreckung von Bauteilen aus verschiedenen Einsatzstählen. HTM Härtereitechnische Mitteilungen, 60, 248-254.

12. Captec Entreprise (2012) Heat Flux Sensor Series.

13. VDI-Gesellschaft Verfahrenstechnik und Chemieingenieurwesen (2010) VDI Heat Atlas. Springer, Berlin.

14. Everett, K.N., Gerner, A. and Durston, D.A. (1983) Seven-Hole Cone Probes for High Angle Flow Measurement Theory and Calibration. AIAA Journal, 21, 992-998.

15. Pellegrino, G., Chaffotte, F., Douce, J.F., Denis, S., Bellot, J.P. and Lamesle, P. (2005) Efficient Numerical Simulation Techniques for High Pressure Gas Quenching. Heat Treating: Proceedings of the 23rd Heat Treating Society Confe- rence, Pittsburgh, 25-28 September 2005, 320-328.

16. Schmidt, R.R., Fritsching, U. and Lübben, T. (2009) Prozessmöglichkeiten zur gezielten Einstellung des Gasab- schreckens. HTM Journal of Heat Treatment and Materials, 64, 351-363.

17. Menter, F., Ferreira, J.C., Esch, T., Konno, B. and Germany, A.C. (2003) The SST Turbulence Model with Improved Wall Treatment for Heat Transfer Predictions in Gas Turbines. Proceedings of the International Gas Turbine Congress, Tokyo, 2-7 November 2003.

18. Menter, F.R., Kuntz, M. and Langtry, R. (2003) Ten Years of Industrial Experience with the SST Turbulence Model. Turbulence Heat and Mass Transfer, 4, 625-632.

19. Mangani, L. and Bianchini, C. (2007) Heat Transfer Applications in Turbomachinery. Proceedings of the OpenFOAM International Conference, London.

20. von Starck, A., Mühlbauer, A. and Kramer, C. (2005) Handbook of Thermoprocessing Technologies: Fundamentals, Processes, Components, Safety. Vulkan-Verlag GmbH, Essen.

21. Heuer, V., Faron, D.R., Bolton, D., Lifshits, M. and Löser, K. (2013) Distortion Control of Transmission Components by Optimized High Pressure Gas Quenching. Journal of Materials Engineering and Performance, 22, 1833-1838.

22. Stratton, P.F. and Ho, D. (2000) Individual Component Gas Quenching. In: 5th ASM Heat Treatment and Surface Engineering Conference in Europe and the 3rd International Conference on Heat Treatment with Atmospheres, Gothen- burg, 7-9 June 2000, 367-375.

23. Ward-Smith, A.J., Lane, D.L., Reynolds, A.J., Sahin, B. and Shawe, D.J. (1991) Flow Regimes in Wide-Angle Screened Diffusers. International Journal of Mechanical Sciences, 33, 41-54.

7

A REVIEW OF THE MODELLING OF THERMALLY INTERACTING MULTIPLE BOREHOLES

Seama Koohi-Fayegh and Marc A. Rosen *

Faculty of Engineering and Applied Science, University of Ontario Institute of Technology, 2000 Simcoe Street North, Oshawa, ON L1H 7K4, Canada

ABSTRACT

Much attention is now focused on utilizing ground heat pumps for heating and cooling buildings, as well as water heating, refrigeration and other thermal tasks. Modeling such systems is important for understanding, designing and optimizing their performance and characteristics. Several heat transfer models exist for ground heat exchangers. In this review article, challenges of modelling heat transfer in vertical heat exchangers are described, some analytical and numerical models are reviewed and compared, recent related developments are described and the importance of modelling these systems is discussed from a variety of aspects, such as sustainability of geothermal systems or their potential impacts on the ecosystems nearby.

KEYWORDS

geothermal energy; modeling vertical ground heat exchangers

1. INTRODUCTION

Measurements show that, below a certain depth in the ground, the temperature fluctuations observed near the surface of the ground diminish, and the temperature remains relatively constant (e.g., at about 6–42 °C in various states in the US) throughout the year [1]. This is due to the high thermal inertia of soil and the time lag between the temperature fluctuations at the surface and their effect deeper in the ground.

Below a certain depth, therefore, the ground generally remains warmer than the outside air in winter and cooler in summer. The relatively cool ground may be used as a sink in summer to store the extracted heat from a conditioned space via a ground heat pump (GHP). In winter, the process may be reversed, and the heat pump can extract heat from the relatively warm ground and transport it into the conditioned space. Compared to a conventional air source heat pump (ASHP), which circulates outdoor air to exchange heat, a ground heat pump exchanges heat by circulating a fluid in the ground. The ground has a lower temperature than the outdoor air in the cooling mode and a higher temperature than the outdoor air in the heating mode. Consequently, the temperature lift across a GHP is lower than that of an air source heat pump for both heating and cooling. Thus, the efficiency of the heat pump, which depends directly on the temperature difference between the circulating fluid and the room, is enhanced for a GHP. Therefore, due to concern about greenhouse gas emissions and high energy prices, the placement of heat loops in the ground is an increasingly common practice for heating and cooling residential, commercial, institutional, recreational and industrial structures. Low temperature geothermal energy has the potential to contribute significantly to mitigating both of these problems.

A geothermal heating and cooling system consists of three main components: a ground heat pump, a ground heat exchanger (GHE) and a distribution system, such as air ducts. GHEs are commonly classified as open loop (groundwater heat pump (GWHP)) or closed loop (ground coupled heat pump (GCHP)), with a third category for those not belonging to either. In an open loop system, ground water from a water-bearing layer is pumped from an aquifer through one well, passed through the heat pump where heat is added to or extracted from a heat carrier and then discharged either onto the surface or to another well in the aquifer. Because the system water supply and discharge are not connected, the loop is "open" [1]. In a similar way, open loop systems can be installed to preheat or pre-cool ambient air flowing through tubes buried in the ground. The air is then heated or cooled by a conventional air conditioning unit before entering the building. A closed loop system uses continuous underground pipe loops placed horizontally or vertically in the ground with both ends of the pipe system connected to the heat pump. In a horizontal GHE, a number of plastic pipes are connected either in series or in parallel in a horizontal trench. The numbers of pipes and trenches installed vary depending on the system capacity and thermal properties of geological formations. This type of GHE is usually most economic when adequate land space is available. A horizontal GHE is usually placed at a depth of 1–2 m in the ground and is typically 35–60 m-long per kW of heating or cooling capacity [1].

Various models have been reported for heat transfer in BHEs that are mostly used in design of BHEs, including sizing borehole depth and determining borehole numbers and analysis of *in situ* ground thermal conductivity test data. A secondary objective of modelling low-temperature geothermal systems is to investigate their sustainability, their potential environmental impact and to provide guidance in regulating the installation of these systems. While the use of geothermal systems is widespread, having had a revival in the

1980s and recently, both the sustainability and impact of these systems on the environment and on their neighbour systems are now being questioned. Due to their efficiency, the use of geothermal energy should be encouraged. However, little research is available to guide regulatory agencies and industry towards designs and installations that maximize their sustainability and minimize possible environmental impacts.

2. OBJECTIVES OF GHE MODELING

The main objective of modelling GHEs is to determine the temperature of the fluid running in the U-tube that exchanges heat between the soil and the heat pump. Under certain operating conditions and building loads, the size and number of the GHE needed to deliver or extract heat to/from the ground is determined according to the acceptable range of the temperature variations of the running fluid. Modelling and simulation of the heat exchange function in GHEs can also be used to evaluate the temperature rise in the soil surrounding these systems and the migration of thermal plumes away from them. This knowledge will guide proper site characterization, system design, construction and operation so that these systems are sustainable and impact the environment as well as other neighbouring systems as little as reasonably possible.

2.1. Environmental Impacts

Similar to most human activities, studies show the potential of geothermal heat exchangers for causing environmental impacts. While little research has been done regarding the impact of geothermal systems on the local environment, research on the movement of thermal plumes shows the potential for impact. Migration of thermal plumes away from these systems and changes in temperature from either closed or open loop systems or due to changes in ground water flow patterns from open-loop systems may cause undesirable temperature rises in nearby temperature-sensitive ecosystems where small temperature differences are important. For example, temperature disturbances in the ground caused by the operation of geothermal systems may result in disruption to sensitive life stages of aquatic organisms. Similar environmental effects are observed for heat loop and waterline projects (rivers and lakes) [2]. Markle and Schincariol [3] investigate the potential thermal impacts from below-water-table aggregate extraction on a cool-water stream in Southwestern Ontario, Canada which supports Brook trout and cool-water micro-invertebrates. They demonstrate the persistence of thermal plumes (persisting in an aquifer for 11 months and migration up to 250 m down gradient) and the sensitivity of the aquatic environment to very small temperature perturbations. Their results show that there is a surprisingly narrow range for spawning in cold water streams. They need to be cooled in the summer and warmed in the winter by the groundwater flow. Once the ground water temperature is affected due to the performance of GHEs, it can negatively affect the temperature of the cold water streams, making these sites unsuitable for spawning. A study on the effects of thermal fluctuation on the microorganisms in the aquifers of the geothermal well field shows increases in total microbial number in aquifer samples, which correlated with the increase in temperature in the geothermal well field [4]. Moreover, counts of cultured bacteria suggested that even when no significant differences in total bacterial number were observed, there may have been changes in the types of microorganisms present in the aquifers of the geothermal well field.

What is unknown at this point is whether the environmental impacts of geothermal systems are acceptable considering the fact that they can reduce fossil fuel consumption and, therefore, lower greenhouse gas emissions and if geothermal systems can be developed in a manner that has reasonably small potential for impacting the environment.

2.2. Sustainability

The sustainability of geothermal heat pump systems at their design efficiency is now being questioned due to 'thermal pollution' from the system itself, adjacent systems or the urban environment. Studies from Manitoba, where the carbonate rock aquifer beneath Winnipeg has been exploited in thermal applications since 1965, indicate that in many cases these systems are not sustainable or not sustainable at the design efficiency [5,6,7]. In an area of the Carbonate Rock aquifer beneath Winnipeg in Manitoba, Canada, there are four systems that utilize groundwater for cooling purposes that are closely spaced. Temperatures at the production well have risen as a result of breakthrough of injected water. The results of numerical modeling also indicate that interference effects are present in three of the four systems examined in this study [6]. The influence of these systems on each other implies that these systems have a spacing that is smaller than the optimum for such systems, and indicates that there is a limit to the density of development that can occur in a given aquifer.

In heating or cooling dominated climates, an annual energy imbalance is placed on the ground loop due to heating, cooling and hot water production. For example, Manitoba has a heating dominated climate and there are concerns regarding the long-term thermal performance of the ground loop. Long-term thermal performance of such ground loop systems with imbalanced energy input and outputs in the ground may result in large temperature rises in the region that the loop is installed. Thermal imbalances could cause significant issues with a heat pump's long-term sustainable performance if not properly considered at the design phase [8].

2.3. Thermal Interaction

Thermal disturbances in the soil associated with GHEs are likely to extend beyond property boundaries and affect adjacent properties. Therefore, with increasing interest in installing such systems in the ground and their potential dense population in coming years, procedures and regulations need to be implemented to prevent disputes between neighbours with potentially interacting systems and their possible negative effects on the performance of existing nearby systems. As stated by Ferguson [9], an analogy exists between ground water and heat flow in the ground and, in many ways, the problem of distributing subsurface energy rights is similar to water rights.

Careful management of geothermal developments to ensure fair access to the subsurface for thermal applications is likely needed. This will require a greater understanding of subsurface heat flow and input from the scientific and technical communities. These concerns have not been well addressed in all cases. Research is needed to allow the investigation of system performance and environmental impact in an integrated manner, so that the best way of utilizing geothermal systems in an environmentally sensitive and sustainable manner can be determined.

3. MODELING GROUND HEAT EXCHANGERS

The heat transfer modelling in GHEs is complicated since their study involves transient effects in a time range of months or even years. Because of the complexities of this problem and its long time scale, the heat transfer in GHEs is usually analyzed in two separated regions (Figure 1): the region inside the borehole containing the U-tubes and the grout and the soil region surrounding the borehole.

The transient borehole wall temperature is important for engineering applications and system simulation. It can be determined by modeling the region outside the borehole by various methods, such as the line source theory. Based on the borehole wall temperature, the fluid inlet and outlet temperatures can be evaluated by a heat transfer analysis inside the borehole. In other words, the regions inside and outside borehole are coupled by the temperature of borehole wall. The heat pump model can utilize the fluid inlet and outlet temperatures for the GHE, and, accordingly, the dynamic simulation and optimization design for a GCHP system can be implemented. This is the basic idea behind the development of the two-region vertical GHE model. Based on how heat transfer from the circulating fluid to the surrounding soil is simulated, these methods can be divided into analytical and numerical. Semi-analytical techniques have also been utilized to describe temperature distributions inside the boreholes as well as outside of them. In these methods, usually analytical solutions are combined with the numerical methods or analytical expressions requiring numerical integrations for evaluation of temperature rise and heat flows. Eskilson and Claesson [10] model the interaction between the convective heat flow in a heat exchanger and heat conduction in the soil using a combination of Laplace transforms and the finite difference method.

3.1. Analytical Approach

In the analytical approaches, heat transfer inside the borehole wall, *i.e.*, from the circulating fluid to the borehole wall, is usually modelled separately than the heat transfer outside the borehole wall, *i.e.*, from the borehole wall to the surrounding soil.

3.1.1. Heat Transfer inside the Borehole

The thermal analysis in the borehole seeks to define the inlet and outlet temperatures of the circulating fluid according to borehole wall temperature, its heat flow and the thermal resistance inside the borehole. The latter quantity is determined by thermal properties of the grouting material, the arrangement of flow channels and the convective heat transfer in the tubes. If the thermal resistance between the borehole wall and inner fluid is determined, the GHE fluid temperature can be calculated. Neglecting natural convection, moisture flow and freezing, the borehole thermal resistance can be calculated assuming steady-state heat conduction in the region between the circulating fluids and a cylinder around the borehole when the running time is greater than the critical time, that is *Fo* > 5, where *Fo* is the Fourier number ($Fo = \alpha t / r_b^2$), and the impact of thermal capacity of objects inside the borehole can be neglected [11]. Such simplification has been proved approximate and convenient for most engineering analyses dealing with responses of more than a few hours [12].

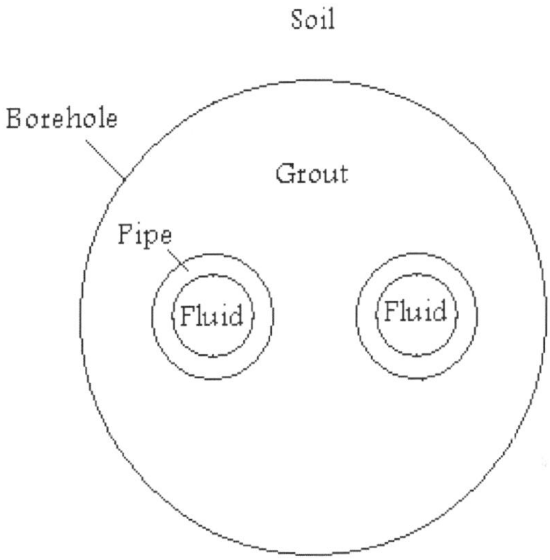

Figure 1. Cross-section of a vertical ground heat exchanger (GHE). The fluid is ascending in one pipe and descending in the other.

In all analytical models for inside of borehole, the axial heat flows in the grout and pipe walls are considered negligible, as the borehole dimensional scale is small compared with the infinite extent of the ground beyond the borehole [13]. However, the fluid circulating through different legs of the U-tube exchanges heat with the surrounding ground and is of varying temperature along the tube. In particular, when the flow rate is low, there is a bigger temperature difference between the upward and downward channels which may result in thermal interference between the two channels and degrades efficiency of the GHE. Due to the U-tube structure, the heat conduction in the cross section is clearly two-dimensional, and the variation of the fluid temperature along the borehole length is in the third dimension.

In some models, such as the Equivalent Diameter method [14,15,16] and the Shape Factor method [17], the U-tube is conceived as a single pipe, and heat transfer in the borehole is approximated as a steady-state one-dimensional process. This oversimplified one-dimensional model is not capable of evaluating thermal interference among borehole legs, analyzing dynamic responses within a few hours, or the axial temperature gradient along the borehole. In the two-dimensional model [18], the effect of U-tube placement is accounted for in the heat conduction problem. The temperature of the fluid in the U-tube is defined by superposing two separate temperature responses caused by the heat fluxes per unit length from the two pipes of the U-tube. A quasi-three-dimensional model was proposed by Zeng *et al.* [19,20], taking into account the fluid axial convective heat transfer and thermal "short-circuiting" among U-tube legs. At the instance of symmetric disposal of the U-tube inside the borehole, the temperature profiles in the two pipes are reduced as:

$$\Theta_1(Z)=\cosh(\beta Z)-\frac{1}{\sqrt{1-P^2}}\left[1-P\frac{\cosh(\beta)-\sqrt{\frac{1-P}{1+P}}\sinh(\beta Z)}{\cosh(\beta)+\sqrt{\frac{1-P}{1+P}}\sinh(\beta)}\right]\sinh(\beta Z)$$

$$\Theta_2(Z)=\frac{\cosh(\beta)-\sqrt{\frac{1-P}{1+P}}\sinh(\beta)}{\cosh(\beta)+\sqrt{\frac{1-P}{1+P}}\sinh(\beta)}\cosh(\beta Z)+\frac{1}{\sqrt{1-P^2}}\left[\frac{\cosh(\beta)-\sqrt{\frac{1-P}{1+P}}\sinh(\beta)}{\cosh(\beta)+\sqrt{\frac{1-P}{1+P}}\sinh(\beta)}-P\right]\sinh(\beta Z) \tag{1}$$

where the dimensionless parameters are defined as:

$$\Theta=\frac{T_f(z)-T_b}{T_f'-T_b},\quad Z=\frac{z}{H},\quad P=\frac{R_{12}}{R_{11}}$$

$$\beta=\frac{H}{\dot{m}c_p\sqrt{(R_{11}+R_{12})(R_{11}-R_{12})}} \tag{2}$$

where z denotes the direction along the tube, H the borehole length, T_f the circulating fluid temperature, T_b the borehole wall temperature, and T_f' the temperature of the fluid entering the U-tube. Also, R_{11} and R_{22} are the thermal resistances between inlet and outlet legs of the U-tube and the borehole wall, respectively, and R_{12} is the thermal resistance between the inlet and outlet legs of the U-tube (Figure 2). These thermal resistances can be calculated analytically using the Multipole Method [18,21]. It is seen in Equation (1) that the Quasi-3-D model is able to reflect the variation of the temperature of the circulating fluid (T_f) along the tube (Z). Quasi-3-D models are preferred for design and analysis of GHEs, as they provide more accurate information on the heat flows inside the borehole.

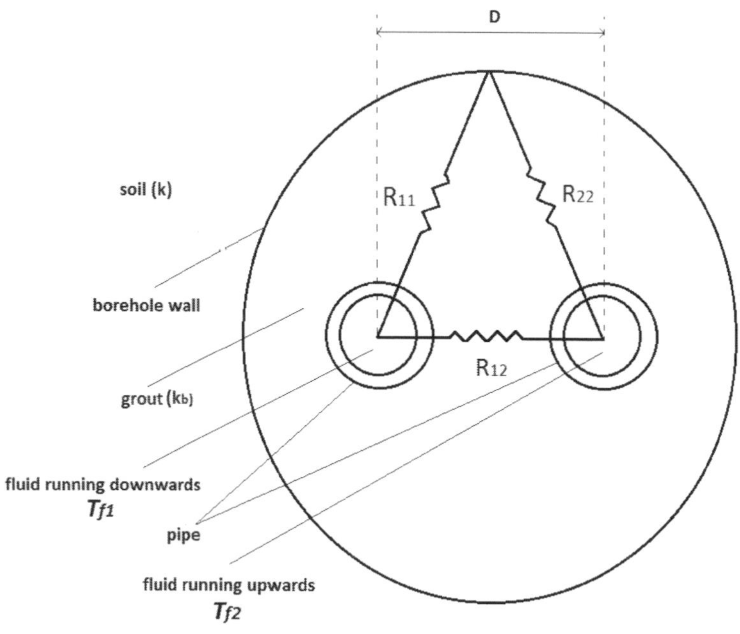

Figure 2. Thermal resistances in the borehole.

Many of the models for heat transfer analysis inside the borehole are summarized in Table 1. Further improvements, such as accounting for thermal capacity of the borehole components to improve the precision of the resistance model, can be made to improve the accuracy of the model inside the borehole [22,23,24].

3.1.2. Heat Transfer outside the Borehole

Several simulation models for the heat transfer outside the borehole are available. The main objective of heat transfer analysis outside the borehole is to determine the transient borehole surface temperature, which is the key to the heat transfer analysis inside the borehole. The models vary in the way the problem of heat conduction in the soil is solved and the way the interference between boreholes is treated.

In the analysis of GHE heat transfer, some complicating factors, such as groundwater movement [25] usually prove to be of minor importance and are analyzed separately. Therefore, the problem of heat transfer outside the borehole becomes a heat conduction problem. The general heat conduction equation in cylindrical coordinates appears in the following form:

$$\frac{\partial^2 T}{\partial r^2}+\frac{1}{r}\frac{\partial T}{\partial r}+\frac{1}{r^2}\frac{\partial^2 T}{\partial \varphi^2}+\frac{\partial^2 T}{\partial z^2}+\frac{\dot{q}_{gen}}{k}=\frac{1}{\alpha}\frac{\partial T}{\partial t} \tag{3}$$

where t denotes the time from the start of operation, α the thermal diffusivity of soil, and T the temperature of the ground. The first two terms on the left side of Equation (3) are the heat flux components in the radial (r) direction, the third and the fourth terms are related to the circumferential (φ) and axial (z) directions, respectively, and the fifth term relates to the heat generated in the control volume. The right side of Equation (3) represents the transient effects of heat conduction.

Table 1. Comparison of various methods in the heat analysis inside the borehole.

	1D (Equivalent diameter) [14]	1D (Shape factor) [17]	2D [18]	Quasi 3D [20]
U-tube disposal	N	Y	Y	Y
Quantitative expressions of the thermal resistance in the cross-section	N	N	Y	Y
Thermal interference	N	N	N	Y
Extinction between the entering and exiting pipes	N	N	N	Y
Axial convection by fluid flow	N	N	N	Y
Axial conduction in grout	N	N	N	N

Unlike the area inside the borehole, heat conduction outside the borehole exhibits transient behavior. As a basic problem, the following assumptions are commonly made:

- The ground is homogeneous in its thermal properties and initial temperature.
- Moisture migration is negligible.
- Thermal contact resistance is negligible between the pipe and grout and between the grout and soil.
- The effect of ground surface is negligible for the initial 5–10 years (depending on the borehole depth).

Due to its minor order, heat transfer in the axial direction along the borehole, which accounts for the heat flux across the ground surface and down to the bottom of the borehole, is considered negligible. This assumption is valid for a length of the borehole distant enough from the borehole top and bottom. Additionally, heat transfer in the circumferential direction is negligible in this model assuming a single borehole. Therefore, the heat transfer is usually modeled with a one-dimensional analysis assuming that the axial and circumferential heat flows are negligible.

The earliest approaches to calculating the heat transfer in the soil surrounding a GHE is Kelvin's line-source model, *i.e.*, the infinite line-source, which uses Fourier's law of heat conduction [26]. In the line-source theory, the borehole is assumed as an infinite line source of heat in the ground, which is regarded as an infinite medium with an initial uniform temperature. This model derives an analytical relation for the temperature excess of the soil, assuming a constant heat flow rate on the borehole wall (here, the line source). Another one-dimensional model based on Fourier's law of heat conduction is the cylindrical source model [27]. In this model, the borehole is assumed to be a cylindrical pipe with infinite length buried in the ground. The governing equation for this model can be solved analytically for either a constant borehole surface temperature or a constant heat transfer rate from the borehole to the ground. At long distances and after a long time, the two models are equivalent. However, the line source model is more popular, since its solution is easier to evaluate, and it is numerically more stable [28]. To make the analytical results obtained by the line source theory for analysis outside of the borehole more accurate and comparable to numerical ones, several studies have focused on improvements [29,30,31,32]. Hellström [18], Kavanaugh [33], Bernier *et al.* [34] and Hikari *et al.* [35] focus on improvements on cylindrical source solution.

In both analytical models of Kelvin's theory and the cylindrical source model, the borehole depth is considered infinite, and the axial heat flow along the borehole depth is assumed negligible. Furthermore, when time tends to infinity, the temperature rise of the Kelvin's theory for an infinite line source tends to infinity, making the infinite model weak for describing heat transfer mechanism in long time steps. Therefore, they can only be used for short time range of operations of GCHP systems. To take into account axial temperature changes for boreholes with finite lengths and in long durations, Eskilson's approach to the problem of determining the temperature distribution around a borehole is based on combination of analytical and numerical solution techniques. Eskilson [36] applies a numerical finite-difference method to the transient radial-axial heat conduction equation for a single borehole. Based on the Eskilson's model [36], Zeng *et al.* [19,37] and Diao *et al.* [38] present an analytical solution to the transient finite line-source problem considering the effects of the finite borehole length and the ground surface as an isothermal boundary. They assume no temperature change on the ground surface (by superimposing an identical mirror borehole above the ground surface with negative strength). With these assumptions, the solution of the temperature excess for a heating rate per length of the constant source (q') is:

$$T(r,z,t)-T_0=\frac{q'}{4k\pi}\int_0^H\left\{\frac{erfc\left(\frac{\sqrt{r^2+(z-h)^2}}{2\sqrt{\alpha t}}\right)}{\sqrt{r^2+(z-h)^2}}-\frac{erfc\left(\frac{\sqrt{r^2+(z+h)^2}}{2\sqrt{\alpha t}}\right)}{\sqrt{r^2+(z-h)^2}}\right\}dh \tag{4}$$

where T is the temperature of the ground at radial distance r, axial distance z, and at time t, T_0 is the initial temperature of the ground, and q' is the heat flow rate per unit length of the borehole. Note that in Equation (4), the heat flow rate per unit length of the borehole is assumed constant along the borehole and steady throughout the operation time. Using this solution as a basic step pulse allows the calculation of temperature response to any load varying with time by considering piecewise constant heat extractions/rejections and superposing them in time as a series of load steps [36]. However, for long time periods, this process becomes computationally intensive. Some authors have focused on introducing efficient algorithms that lower the number of time steps [34,39], while others have used the convolution theorem in the frequency domain using fast Fourier transform to lower the computational burden of hourly temperature evaluations related to a time-varying heat load [28].

In the analytical models presented above, a number of assumptions are employed in order to simplify the complicated governing equations. In time varying heat transfer rates and the influence of surrounding boreholes on both long and short time scales, analytical methods are not as suitable as numerical methods. However, due to their much shorter computation times, they are still used widely in designing GHEs.

3.2. Numerical Models

System simulation models require the ability to operate at short time scales, often less than one minute. Therefore, the dynamic response of the grout material inside the borehole should be considered. This is possible when the model is solved using numerical techniques.

Numerical methods have also been used extensively for evaluating the heat conduction inside the borehole and the soil surrounding it [11,40,41,42,43,44,45,46,47,48,49,50]. In addition to short time step solution, numerical techniques are advantages over the available analytical ones due to accounting for all the terms in the conduction equation [Equation (3)] and the ability to apply transient boundary conditions on the model such as the surface, borehole wall and the inlet temperature boundaries. Model additions, such as accounting for moisture migration in the soil and ground stratification, are easier to be made in a numerical model than the analytical ones.

One of the disadvantages of numerical approaches is their computation time for long-term system performance. The diameters of the U-tubes in the borehole are fairly small, on the order of 10^{-2} m, while the size of the solution domain, which depends on the duration of system operation and its heating/cooling load, is approximately on an order of 10 m, making the domain extremely disproportionate. As a result, a large number of mesh elements is required for simulation of a single borehole and its surrounding soil. To achieve an inaccuracy of 2% or less for the steady state heat transfer analysis of boreholes, a minimum number of approximately 18 elements describing any circular shape of a horizontal cross section is needed [51]. In modelling the soil surrounding the borehole, a domain of a certain size can work well for one model, while it can be too small for another model requiring more boreholes, longer system performance durations or higher heating injection/removal rates. At the outer edge of the domain, a constant far-field temperature condition equal to the initial temperature is often applied. The sensitivity of the solution results to this boundary should always be examined and avoided by increasing the size of the domain. In three-dimensional modeling of a borehole system with typical flow velocities, a vertical element size of 2 m or less should often be applied to avoid inaccuracies of greater than 2% [51].

The temperature gradient in the domain between the borehole wall and the far field changes gradually from large to small. Therefore, to reduce computer memory and computational time, the size of the mesh cells is often chosen based on this gradual change. Furthermore, the symmetry about the GHE can be used to save computation time. In such cases, the symmetric portion of the solution domain is replaced by an adiabatic wall boundary condition on the symmetry line. Applying all these techniques, a three-dimensional 15 m × 15 m × 60 m domain may require mesh sizes of the order of 1,000,000 elements to simulate multiple boreholes of 50 m length.

Due to these limitations, several available models are limited to a two-dimensional (2D) description of the domain [41,50]. Factors such as vertical heat transport in and outside the borehole, different ground layers, the vertical gradient of the undisturbed ground temperature, the transient fluid transport inside the tubes, the thermal short-circuiting between the upward and downward tubes, and the correct boundary conditions at the upper and lower boundaries are generally ignored in the 2D models. Some models describe the system in a three-dimensional domain by reducing the number of mesh elements, but this approach can result in lower accuracy. There are also several numerical techniques available in different numerical approaches which can be used to reduce the computation time [23,24,51].

To evaluate the long-term temperature response in the soil surrounding multiple borehole systems, a numerical finite volume method in a two-dimensional meshed domain is used previously [52]. The effect of installing GHEs in the ground and the temperature rise in the soil over the long-term, for a period of five years, is considered. A transient periodic heat flux is assumed on the borehole wall reflecting the annual variation of heat storage/removal in the ground. When monthly bin weather information, building needs and heat pump performance and efficiency data are available, this periodic heat flux can be calculated and used as a heat boundary condition in the numerical model. It is assumed that flow rate and the inlet temperature of the circulating fluid running in the U-tube inside the borehole will be adjusted according to the building heating needs. The five-year simulation of the system shows that for a system that has a balanced heat injection and extraction into the soil, if the borehole spacing and the heat injection/extraction rate are designed within acceptable limits, there will not be any temperature increase or decrease in the soil surrounding the system. Any temperature rise or decrease in the soil surrounding the GHE that is noticed after the first year operation needs to be compensated for during the second year operation so that the system can operate sustainably.

One limitation in the previous studies [52] is using a heat flux boundary condition on the borehole wall. As mentioned previously, using a heat flux boundary condition can cause the temperature of the ground to rise infinitely without a stop in system operation. In reality, if the temperature of the soil surrounding a borehole becomes close to or higher than the inlet temperature of the circulating fluid exiting the heat pump, the system will not be able to deliver the desired heat to the ground and will automatically stop operating until the heat around it is dissipated away and the temperature drops to a lower value. In order to overcome such a limitation, the periodic heat boundary on the borehole wall can be replaced with a temperature boundary or the heat boundary can be updated at short time steps with respect to the soil temperature. This is possible if the heat transfer model for outside of the borehole is coupled to the model inside the borehole.

3.3. Some Modeling Limitations

One limitation in most of the previous studies is the assumption of uniform heat input along the borehole length to the ground, either when the borehole is assumed as a cylinder or a line source of heat. In order to determine the borehole heat delivery/removal profile along the borehole, the heat transfer model outside the borehole should be coupled to the one inside the borehole. In a recent study, Koohi-Fayegh and Rosen [53] use the analytical quasi-three-dimensional solution to the heat transfer problem of the U-tube configuration inside the borehole. This model evaluates the temperature of the circulating fluid along the borehole length and is used in the model for outside the borehole to calculate the heat delivery/removal along the borehole caused by the temperature difference between the circulating fluid and the borehole wall temperature. The heat delivery/removal is implemented as the heat boundary condition in the analytical line source with finite length as well as in a three-dimensional finite volume model [54, 55]. The results show that due to the higher heating strength at the top end of the boreholes (about 3% total length), the possibility of thermal interaction at the top of the borehole is at its highest, and it decreases along the borehole length as the heat flux from the borehole wall into the soil decreases. Therefore, with the objective of limiting boreholes' operations and sizes in order to prevent their thermal interaction, the top length of the boreholes is the critical area. It can be concluded that using a uniform heat flux along the borehole is only accurate for the middle length of the boreholes, and moving any further to the top or bottom of the borehole, the temperature rise evaluations become relatively inaccurate.

Another limitation in the previous studies is the assumption of steady borehole wall temperature during system operation. When calculating the heat input to the ground, it becomes clear that it varies with the borehole wall temperature. Although the soil temperature at the borehole wall rises as the system operates, it is often assumed that the soil temperature at the borehole wall is constant throughout the operation period. This assumption ignores the drop in heat injection strength when the borehole wall temperature increases and, therefore, underestimates the inlet temperature of the circulating fluid that is required to meet the heat injection needs of the system. Yang *et al.* [56] propose and develop an updated two-region vertical U-tube GHE analytical model coupling two solutions for inside and outside the borehole with transient temperature of borehole wall. In the fewer cases of multiple boreholes, superposition of the temperature excesses resulted from individual boreholes seems to be the most popular solution in analytical approaches. In numerical approaches, the boundary condition that plays the role of heat delivery/removal is a heat boundary type that, regardless of being constant or variable based on the building needs, does not reflect the drop in the heat injection/removal strength when temperature of the soil around the borehole increases/decreases by its own performance or another nearby system's performance. This assumption forces the system to deliver a desired amount of heat to the ground regardless of the ground temperature. In reality, the amount of heat delivered to the ground is driven by the temperature difference between the circulating fluid and the ground temperature. In some cases, the assumption of constant borehole wall temperature is acceptable considering how the conduction problem is simplified. However, when determining how thermal interaction between two operating GHEs can affect their performance, the effect of the transient borehole wall temperature on their heat delivery strength and inlet fluid temperature becomes a very important factor. Therefore, this model is only able to illustrate the variation of the heat delivery/removal strength when the heat flow rate is low and the temperature changes at the borehole wall are small.

In order to account for higher heat flow rates or thermal interaction between multiple boreholes, the model should be modified to include the transient value of borehole wall temperature. Thus, the non-uniform heat flow rate along the borehole wall becomes transient as well. A model is needed that is able to not only estimate how heat flows in the region surrounding GHEs, but also how a temperature rise in the soil surrounding a borehole caused by the system itself or a neighboring geothermal system can interfere with its heat delivery strength.

3.4. Other Modeling Aspects

Performing energy and moisture balances at the ground surface involves very complex processes, taking into account solar radiation, cloud cover, surface albedo, ambient air temperature and relative humidity, rainfall, snow cover, wind speed and evapotranspiration. Such details provide a proper account of the renewable energy resource. However, due to the complexity of adding all the above heat fluxes in a numerical model, some studies assume the ground surface temperature variation at the ground surface to take the form of a sine wave or Fourier series [57,58,59,60], while, in most analytical approaches, the ground surface boundary is assumed to have a constant temperature equal to the soil temperature deep in the ground (by superimposing an identical mirror borehole above the ground surface with negative strength). Moreover, some studies simplify the problem further and assume an adiabatic boundary condition at the ground surface.

Neglecting the existence of moisture in the soil, the heat flux is described via the conduction heat flow. The coupled heat and moisture flow in a soil system is described with a thermal energy balance coupled with a mass balance. This adds to the complication of the problem since the complete model contains a set of transient simultaneous partial differential equations with many soil parameters that are not readily available. Research shows that the effects of moisture migration are not significant to the operation of a vertical GHE; it is expected that these effects are more pronounced with a horizontal GHE. This is because natural variations of temperature and moisture near the ground surface and operation of the HGHE may create a potentially greater moisture movement. During the cooling season, migration of soil moisture away from the GHE may lead to a drastic drop in soil thermal conductivity and consequently a significantly reduced heat transfer, which has a devastating effect on GHE performance. Therefore, although moisture migration effects can be neglected in early stages of design or conceptual development, not considering them in long-term operation of GCHP systems makes it impossible to assess the performance and potential failure of these systems [61]. Some studies focus on the thermal interaction between the circulating fluid and soil, taking into account heat flow with moisture transfer in the soil [40,43,61,62].

A further complication in the design of ground-coupled heat pump systems is the presence of groundwater. Due to the difficulties encountered both in modeling and computing the convective heat transfer and in learning about the actual groundwater flow in engineering practice, each of the methods presented in the previous sections is based on Fourier's law of heat conduction and neglect the effects of groundwater flow in carrying away heat. Where groundwater is present, flow will occur in response to hydraulic gradients, and the physical process affecting heat transfer in the ground is inherently a coupled one of heat diffusion (conduction) and heat advection by moving groundwater [63]. Similar to the models discussed in the previous sections, here as well, the objective is to evaluate the temperature response in the soil surrounding GHEs. The effect of ground water is analyzed using numerical [26,64,65,66] and analytical approaches [63,67].

4. CONCLUSIONS

An assessment of the available analytical models demonstrates that they are not capable of estimating the heat delivery/removal strength when the soil surrounding them experiences a temperature rise. In the current study, it is shown that the effect of the temperature rise in the soil surrounding boreholes is not negligible. The distance between two boreholes or two systems of boreholes, the heat flux from the borehole wall and the time of system operation all affect directly the amount of thermal interaction between the systems. However, the effect of these parameters on system operation and heat delivery/removal rate can only be studied in models that account for the change in the borehole wall temperature.

As mentioned in the previous section, in order to account for the sustainability of the system and heat pump efficiency when thermal interaction among boreholes occur, it is important to develop and utilize models that account for the drop in heat delivery strength when the borehole wall temperature increases during the operation time or by another nearby operating system. As a result, the inlet temperature of the circulating fluid needs to be adjusted to a higher/lower one to maintain the heat delivery/removal needs of the system. This analysis is important since ground heat exchangers are coupled to a heat pump that can only work within a certain temperature lift and inlet and outlet circulating temperature ranges. If a system is able to deliver a certain amount of heat to the ground, the increase in the inlet circulating temperature due to temperature rise in the soil caused by a nearby system reflects how thermal interaction affects the sustainability of the system. Furthermore, simulation of heat exchange processes within the system and surrounding environment through local scale assessment, simulation of migration of thermal plumes into the hydrogeological environment through intermediate and regional scale assessment will help gain an estimation of ecological impacts.

ACKNOWLEDGMENTS

The support provided by the Ontario Ministry of Environment through its Best in Science program is gratefully acknowledged.

REFERENCES

1. Geothermal Heat Pump Consortium. Available online: http:// energy.nstl. gov.cn/MirrorResources/902/index.html/ (accessed on 1 November 2011).

2. Fisheries and Oceans Canada, Ontario-Great Lakes Area (DFO-OGLA). Fish Habitat and Fluctuating Water Levels on the Great Lakes. Available online: http://www.dfo-mpo.gc.ca/ regions/ central/pub/factsheets-feuilletsinfos-on/t2-eng.htm (accessed on 1 November 2011).

3. Markle, J.M.; Schincariol, R.A. Thermal plume transport from sand and gravel pits: Potential thermal impacts on cool water streams. *Hydrology* **2007**, *338*, 174–195.

4. York, K.P.; Sarwar Jahangir, Z.M.G.; Solomon, T.; Stafford, L. Effects of a Large Scale Geothermal Heat Pump Installation on Aquifer Microbiota. In Proceedings of the Second International Conference on Geothermal Heat Pump systems at Richard Stockton College, Pomona, NJ, USA, 16–17 March 1998.

5. Ferguson, G.; Woodbury, A.D. Thermal sustainability of groundwater-source cooling in Winnipeg, Manitoba. *Can. Geothechnical J.* **2005**, *42*, 1290–1301.

6. Ferguson, G.; Woodbury, A.D. Observed thermal pollution and and post-development simulations of low-temperature geothermal systems in Winnipeg, Canada. *Hydrogeology* **2006**, *14*, 1206–1215.

7. Younger, P.L. Ground-coupled heating-cooling systems in urban areas: How sustainable are they? *Bull. Sci. Technol. Soc.* **2008**, *28*, 174–182.

8. Andrushuk, R.; Merkel, P. Performance of ground source heat pumps in Manitoba. Available online: http://www.hydro.mb.ca/regulatory_affairs/ electric/gra_2012_2013/Appendix_38.pdf (accessed on 1 November 2011).

9. Ferguson, G. Unfinished business in geothermal energy. *Ground Water* **2009**, *47*, 167–167.

10. Eskilson, P.; Claesson, J. Simulation model for thermally interacting heat extraction boreholes. *Numer. Heat Transf.* **1988**, *13*, 149–165.

11. Jun, L.; Xu, Z.; Jun, G.; Jie, Y. Evaluation of heat exchange rate of GHE in geothermal heat pump systems. *Renew. Energy* **2009**, *34*, 2898–2904.

12. Yavuzturk, C. Modeling of vertical ground loop heat exchangers for ground source heat pump systems. Ph.D. thesis, Oklahoma State University, Oklahoma, OK, USA, 1999.

13. Bose, J.E.; Parker, J.D.; McQuiston, F.C. *Design/Data Manual for Closed-Loop Ground Coupled Heat Pump Systems*; Oklahoma State University for ASHRAE: Stillwater, OK, USA, 1985.

14. Claesson, J.; Dunand, A. *Heat Extraction from the Ground by Horizontal Pipes: A Mathematical Analysis*; Swedish Council for Building Research: Stockholm, Sweden, 1983.

15. Gu, Y.; O'Neal, D.L. Development of an equivalent diameter expression for vertical U-tube used in ground-coupled heat pumps. *ASHRAE Trans.* **1998**, *104*, 347–355.

16. Gu, Y.; O'Neal, D.L. Modeling the effect of backfills on U-tube ground coil performance. *ASHRAE Trans.* **1998**, *104*, 356–365.

17. Paul, N.D. The effect of grout conductivity on vertical heat exchanger design and performance. Master Thesis, South Dakota State University, Madison, SD, USA, 1996.

18. Hellström, G. Ground heat storage: Thermal analyses of duct storage systems. Ph.D. thesis, Department of Mathematical Physics, University of Lund, Lund, Sweden, 1991.

19. Zeng, H.Y.; Diao, N.R.; Fang, Z. Efficiency of vertical geothermal heat exchangers in ground source heat pump systems. *J. Therm. Sci.* **2003**, *12*, 77–81.

20. Zeng, H.Y.; Diao, N.R.; Fang, Z. Heat transfer analysis of boreholes in vertical ground heat exchangers. *Int. J. Heat Mass Transf.* **2003**, *46*, 4467–4481.

21. Claesson, J.; Hellström, G. Multipole method to calculate borehole thermal resistances in a borehole heat exchanger.*HVAC&R Res.* **2011**, *17*, 895–911.

22. Bauer, D.; Heidemann, W.; Muller-Steinhagen, H.; Diersch, H.-J.G. Thermal resistance and capacity models for borehole heat exchangers. *Int. J. Energy Res.* **2011**, *35*, 312–320.

23. Al-Khoury, R.; Bonnier, P.G.; Brinkgreve, R.B.J. Efficient finite element formulation for geothermal heating systems, Part I: Steady state. *Int. J. Numer. Methods Eng.* **2005**, *63*, 988–1013.

24. Al-Khoury, R.; Bonnier, P.G. Efficient finite element formulation for geothermal heating systems, Part II: Transient.*Int. J. Numer. Methods Eng.* **2006**, *67*, 725–745.

25. Chiasson, A.D.; Rees, S.J.; Spitler, J.D. A preliminary assessment of the effects of groundwater flow on closed-loop ground-source heat pump systems. *ASHRAE Trans.* **2000**, *106*, 380–393.

26. Ingersoll, L.R.; Zobel, O.J.; Ingersoll, A.C. *Heat Conduction with Engineering, Geological, and other Applications*, rev. ed.; University of Wisconsin Press: Madison, WI, USA, 1954.

27. Carslaw, H.S.; Jaeger, J.C. *Conduction of Heat in Solids*; Claremore Press: Oxford, UK, 1946.

28. Marcotte, D.; Pasquier, P. Fast fluid and ground temperature computation for geothermal ground-loop heat exchanger systems. *Geothermics* **2008**, *37*, 651–665.

29. Ingersoll, L.R.; Plass, H.J. Theory of the ground pipe heat source for the heat pump. *ASHVE Trans.* **1948**, *47*, 339–348.

30. Hart, D.P.; Couvillion, R. *Earth Coupled Heat Transfer*; National Water Well Association: Dublin, OH, USA, 1986.

31. Lamarche, L.; Beauchamp, B. A new contribution to the finite line source model for geothermal boreholes. *Energy Build.* **2007**, *39*, 188–198.

32. Cui, P.; Yang, H.; Fang, Z.H. Heat transfer analysis of ground heat exchangers with inclined boreholes. *Appl. Therm. Eng.* **2006**, *26*, 1169–1175.

33. Kavanaugh, S.P. A design method for commercial ground-coupled heat pumps. *ASHRAE Trans.* **1995**, *101*, 1088–1094.

34. Bernier, M.; Pinel, A.; Labib, P.; Paillot, R. A multiple load aggregation algorithm for annual hourly simulations of GCHP systems. *HVAC&R Res.* **2004**, *10*, 471–487.

35. Hikari, F.; Ryuichi, I.; Takashi, I. Improvements on analytical modeling for vertical U-tube ground heat exchangers.*Geotherm. Resour. Counc. Trans.* **2004**, *28*, 73–77.

36. Eskilson, P. Thermal analysis of heat extraction boreholes. Doctoral thesis, Department of Mathematical Physics, University of Lund, Lund, Sweden, 1987.

37. Zeng, H.Y.; Diao, N.R.; Fang, Z. A finite line-source model for boreholes in geothermal heat exchangers. *Heat Transf. Asian Res.* **2002**, *31*, 558–567.

38. Diao, N.R.; Zeng, H.Y.; Fang, Z.H. Improvement in modeling of heat transfer in vertical ground heat exchangers.*HVAC&R Res.* **2004**, *10*, 459–470.

39. Yavuzturk, C.; Spitler, J. A short time step response factor model for vertical ground loop heat exchangers. *ASHRAE Trans.* **1999**, *105*, 475–485

40. Mei, V.C.; Baxter, V.D. Performance of a ground-coupled heat pump with multiple dissimilar U-tube coils in series.*ASHRAE Trans.* **1986**, *92*, 22–25.

41. Yavuzturk, C.; Spitler, J.D.; Rees, S.J. A transient two-dimensional finite volume model for the simulation of vertical U-tube ground heat exchangers. *ASHRAE Trans.* **1999**, *105*, 465–474.

42. Yavuzturk, C.; Spitler, J.D. Field validation of a short time step model for vertical ground-loop heat exchangers.*ASHRAE Trans.* **2001**, *107*, 617–625.

43. Muraya, N.K. Numerical modeling of the transient thermal interference of vertical U-tube heat exchangers. Ph.D. Thesis, Texas A&M University, College Station, TX, USA, 1995.

44. Kavanaugh, S.P. Simulation and experimental verification of vertical ground coupled heat pump systems. Ph.D. Thesis, Oklahoma State University, Oklahoma, OK, USA, 1985.

45. Rottmayer, S.P.; Beckman, W.A.; Mitchell, J.W. Simulation of a single vertical U-tube ground heat exchanger in an infinite medium. *ASHRAE Trans.* **1997**, *103*, 651–658.

46. Lee, C.K.; Lam, H.N. Computer simulation of borehole ground heat exchangers for geothermal heat pump systems.*Renew. Energy* **2008**, *33*, 1286–1296.

47. Li, Z.; Zheng, M. Development of a numerical model for the simulation of vertical U-tube ground heat exchangers.*Appl. Therm. Eng.* **2009**, *29*, 920–924.

48. He, M.; Rees, S.; Shao, L. Simulation of a Domestic Ground Source Heat Pump System Using a Transient Numerical Borehole Heat Exchanger Model. In Proceedings of 11th International Building Performance Simulation Association Conference, Glasgow, Scotland, 27–30 July 2009; pp. 607–614.

49. Fang, Z.H.; Diao, N.R.; Cui, P. Discontinuous operation of geothermal heat exchangers. *Tsinghua Sci. Technol.* **2002**, *7*, 194–197.

50. Austin, W.A.; Yavuzturk, C.; Spitler, J.D. Development of an in situ system for measuring ground thermal properties.*ASHRAE Trans.* **2000**, *106*, 356–379.

51. Bauer, D.; Heidemann, W.; Diersch, H.-J.G. Transient 3D analysis of borehole heat exchanger modeling. *Geothermics* **2011**, *40*, 250–260.

52. Koohi-Fayegh, S.; Rosen, M.A. Long-term Study of Thermal Interaction of Vertical Ground Heat Exchangers with Seasonal Heat Flux Variation. In Proceedings of 11th International Conference on Sustainable Technologies, Vancouver, BC, Canada, 2–5 September 2012.

53. Koohi-Fayegh, S.; Rosen, M.A. Thermally Interacting Multiple Boreholes with Variable Heating Strength. In Proceedings of eSim Conference, Halifax, NS, Canada, 2–3 May 2012.

54. Koohi-Fayegh, S.; Rosen, M.A. A Numerical Approach to Assessing Thermally Interacting Multiple Boreholes with Variable Heating Strength. In Proceedings of 1st World Sustainability Forum, 1–30 November 2011.

55. Koohi-Fayegh, S.; Rosen, M.A. On thermally interacting multiple boreholes with variable heating strength: Comparison between analytical and numerical approaches. *Sustainability* **2012**, *4*, 1848–1866.

56. Yang, W.; Shi, M.; Liu, G.; Chen, Z. A two-region simulation model of vertical U-tube ground heat exchanger and its experimental verification. *Appl. Energy* **2009**, *86*, 2005–2012.

57. Salah El-Din, M.M. On the heat flow into the ground. *Renew. Energy* **1999**, *18*, 473–490.

58. Mihalakakou, G. On estimating ground surface temperature profiles. *Energy Build.* **2002**, *34*, 251–259.

59. Mihalakakou, G.; Lewis, J.O. The influence of different ground covers on the heating potential of the earth-to-air heat exchangers. *Renew. Energy* **1996**, *7*, 33–46.

60. Jacovides, C.P.; Mihalakakou, G.; Santamouris, M.; Lewis, J.O. On the ground temperature profile for passive cooling applications in buildings. *Solar Energy* **1996**, *57*, 167–175.

61. Leong, W.H.; Tarnawski, V.R. Effects of Simultaneous Heat and Moisture Transfer in Soils on the Performance of a Ground Source Heat Pump System. In Proceedings of ASME-ATI-UIT Conference on Thermal and Environmental Issues in Energy Systems, Sorrento, Italy, 16–19 May 2010.

62. Piechowski, M. Heat and mass transfer model of a ground heat exchanger: Theoretical development. *Int. J. Energy Res.***1999**, *23*, 571–588.

63. Sutton, M.; Nutter, D.; Couvillion, R. A ground resistance for vertical bore heat exchangers with groundwater flow. *J. Energy Resour. Technol.* **2003**, *125*, 183–189.

64. Gehlin, S.E.A.; Hellström, G. Influence on thermal response test by groundwater flow in vertical fractures in hard rock. *Renew. Energy* **2003**, *28*, 2221–2238.

65. Nam, Y.; Ooka, R.; Hwang, S. Development of a numerical model to predict heat exchange rates for a ground-source heat pump system. *Energy Build.* **2008**, *40*, 2133–2140.

66. Hecht-Mendez, J.; Molina-Giraldo, N.; Blum, P.; Bayer, P. Evaluating MT3DMS for heat transport simulation of closed geothermal systems. *Ground Water* **2010**, *48*, 741–756.

67. Diao, N.R.; Li, Q.; Fang, Z.H. Heat transfer in ground heat exchangers with groundwater advection. *Int. J. Therm. Sci.***2004**, *43*, 1203–1211.

8

NUMERICAL MODELING OF NON-SIMILAR MIXED CONVECTION HEAT TRANSFER OVER A STRETCHING SURFACE WITH SLIP CONDITIONS

A. Subba Rao[1*], V. R. Prasad[1], N. Nagendra[1], K. V. N. Murthy[1], N. Bhaskar Reddy[2], O. Anwar Beg[3]

[1]Department of Mathematics, Madanapalle Institute of Technology and Science, Madanapalle, India

[2]Department of Mathematics, Sri Venkateswara University, Tirupathi, India

[3]GortEngovation-Aerospace, Medical and Energy Engineering, Bradford, UK

ABSTRACT

In this paper, the heat transfer effect on the steady boundary layer flow of a Casson fluid past a stretching surface in the presence of slip conditions was analyzed. The stretching surface is maintained at a constant temperature. The boundary layer conservation equations, which are parabolic in nature, are normalized into non-similar form and then solved numerically with the well-tested, efficient, implicit, stable Keller-box finite difference scheme. The resulting equations are solved numerically by using the Kellerbox finite-difference method, and the expressions for velocity and temperature are obtained. They satisfy all imposed initial and boundary conditions and reduce to some well-known solutions for non-Newtonian fluids. Numerical results for velocity, temperature, skin friction and Nusselt number are shown in various graphs and discussed for embedded flow parameters. It is found that both velocity and temperature decrease with an increase of the Casson fluid parameter.

KEYWORDS

Stretching Surface, Non-Newtonian Fluid, Slip Condition, Keller-Box Numerical Method, Heat Transfer, Skin Friction Coefficient

1. INTRODUCTION

Heat transfer in non-Newtonian fluids is an important research area due to its wide applications in food processing, petroleum production, cooling of an infinite metallic plate in a cooling bath and in many industries, for example polymers melt, polymer solutions employed in the plastic processing, a long thread traveling between a feed roll and a wind-up roll etc. Flow in the boundary layer on moving solid surface was historically first investigated by Sakiadis [1] [2] who observed that the boundary layer growth was in the direction of motion of the continuous solid surface and deviates from that of the classical Blasius flow past a flat plate. Erickson et al. [3] extended the Sakiadis problem to include blowing or suction at the moving surface and investigated its effects on the heat and mass transfer in the boundary layer. Bijjanal et al. [4] obtained closed-form similarity solutions for steady two-dimensional incompressible boundary layer flow caused by a stretching sheet with Non-Uniform Heat Source/Sink. Combined forced and free convection in boundary layers adjacent to a continuous horizontal sheet maintained at a constant temperature and moving with a constant velocity it was investigated numerically by Chen and Strobel [5]. Grubka and Bobba [6] have investigated the stretching sheet problem for a surface moving with linear velocity and with a variable surface temperature. Dutta et al. [7] studied numerically the Temperature Field in Flow over a Stretching Surface with Uniform Heat Flux. Chen and Char [8] investigated the effects of variable surface temperature and variable heat flux on the heat transfer characteristics of a linearly stretching sheet subject to blowing or suction. Several excellent studies of stretching flows in materials processing were presented by Karwe and Jaluria [9] [10]. Patil et al. [11] further analysed unsteady two-dimensional mixed convection flow along a vertical semi-infinite power law stretching sheet in a parallel free stream with a power-law temperature distribution. Nath et al. [12] analysed the three-dimensional, time dependent stretch surface flow. Ali and Al-Yousef [13] analysed mixed convection heat transfer from a uniformly stretching vertical surface with power function form for wall temperature. Partha et al. [14] described the effects of viscous dissipation on mixed convection heat transfer from an exponentially stretching surface.

Non-Newtonian transport phenomena arise in many branches of chemical and material processing engineering. Such fluids exhibit shear-stress-strain relationships which diverge significantly from the Newtonian (Navier-Stokes) model. Most non-Newtonian models involve some forms of modification to the momentum conservation equations. These include power-law, and thixotropic and viscoelastic fluids (Schowalter [15]). Such rheological models however cannot simulate the microstructural characteristics of many important liquids including polymer suspensions, liquid crystal melts, physiological fluids, contaminated lubricants, etc. Several fluids in chemical engineering, multiphase mixtures, pharmaceutical formulations, china clay and coal in water, paints, synthetic lubricants, salvia, synovial fluid, jams, soups, jellies, marmalades, sewage sludge etc. are non-Newtonian. The constitutive relations for these kinds of fluids give rise to more complex and higher order equations than the Navier-Stokes equations. Considerable progress has been made on the topic by using different models of non-Newtonian fluids [16] - [25]. Previous studies indicate that not much has been presented yet regarding Casson fluid. This model (Casson [26]; Nakamura et al. [27]; Samir Kumar [28]) in fact is a plastic fluid that exhibits shear thinning characteristics and that quantifies yield stress and high shear viscosity. Casson fluid model is reduced to a Newtonian fluid at a very high wall shear stress, when wall stress is much greater than yield stress. This fluid has good approximations for many substances such as biological materials, foams, molten chocolate, cosmetics, nail polish, some particulate suspensions etc. The boundary layer behaviour of viscoelastic fluid has technical applications in engineering such as glass fibre, paper production, manufacture of foods, the aerodynamic extrusion of plastic sheets, the polymer extrusion in a melt spinning process and many others.

Most of the existing studies on steady boundary layer flow and heat transfer with slip conditions are limited to the non-Newtonian fluid. The considered slip conditions especially are important in the non-Newtonian fluids such as polymer melts which often exhibit wall slip. This motivates us to consider the slip conditions in the present work for non-Newtonian fluids. More exactly, our aim is to investigate steady boundary layer flow and heat transfer of a Casson fluid past a stretching sheet with slip conditions. The equations of the problem are first formulated and then transformed into their dimensionless forms where the Keller box method is applied to find the exact solutions for velocity, temperature, Skin-friction and Nusselt number.

2. MATHEMATICAL ANALYSIS

We consider steady two-dimensional laminar mixed convection heat transfer flow along a stretching surface with partial slip. By applying two equal and opposing forces along the x-axis, the sheet is stretched with a speed proportional to the distance from the fixed origin $x = 0$ as shown in Figure 1. It is also assumed that the external electric field is zero and the electric field due to the polarization of chargers is negligible. The temperature is maintained at prescribed constant value. The fluid properties are assumed to be constant except the density variation in the buoyancy force term.

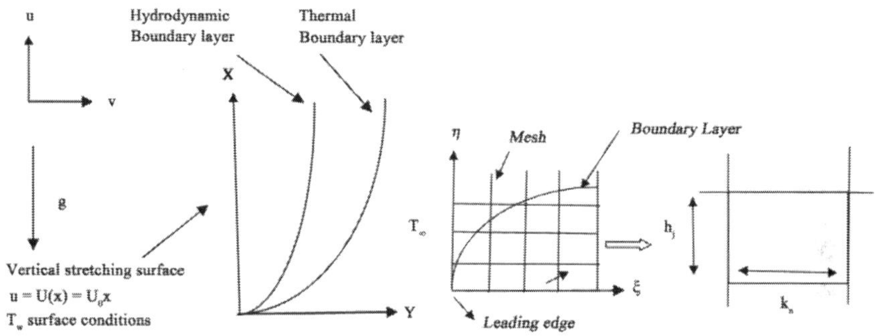

(a) (b)

Figure 1. (a) Physical model and coordinate system; (b) Grid meshing and a Keller box computational cell.

The rheological equation of state for an isotropic flow of Casson fluid is (Nakamura et al. [27]):

$$\tau_{ij} = \begin{cases} 2\left(\mu_B + \dfrac{p_y}{\sqrt{2\pi}}\right)e_{ij}, \pi \geq \pi_c \\ 2\left(\mu_B + \dfrac{p_y}{\sqrt{2\pi_c}}\right)e_{ij}, \pi < \pi_c \end{cases} \tag{1}$$

in which $\pi = e_{ij}e_{ij}$ and e_{ij} represents the $(i,j)^{th}$ component of deformation rate, μ is the dynamic viscosity, π denotes the product of the component of deformation rate with itself, π_c shows a critical value of this product based on the non-Newtonian model, μ_B represents the plastic dynamic viscosity of non-Newtonian fluid and p_y is the yield stress of fluid.

Under the usual Boussinesq and boundary layer approximations, the equations for mass continuity, (continuity/mass conservation) momentum and energy can be written in the following form:

$$\frac{\partial u}{\partial x}+\frac{\partial v}{\partial y}=0 \tag{2}$$

$$u\frac{\partial u}{\partial x}+v\frac{\partial u}{\partial y}=\nu\left(1+\frac{1}{\beta}\right)\frac{\partial^2 u}{\partial y^2}+g\Omega\left(T-T_\infty\right) \tag{3}$$

$$u\frac{\partial T}{\partial x}+v\frac{\partial T}{\partial y}=\alpha\frac{\partial^2 T}{\partial y^2} \tag{4}$$

where u and v are the velocity components in the x- and y-directions, ν is the kinematic viscosity of the conducting fluid, β is the non-Newtonian Casson parameter, α is the thermal diffusivity, T is the temperature respectively.

The boundary conditions are prescribed at the stretching surface and the edge of the boundary layer regime, respectively as follows:

$$\text{At } y=0,\;\; u=U\left(x\right)+N_0\left(1+\frac{1}{\beta}\right)\frac{\partial u}{\partial y},\; v=0,\; T=T_w+K_0\frac{\partial T}{\partial y}$$

$$\text{As } y\to\infty,\;\; u\to 0,\;\; T\to T_\infty \tag{5}$$

where N_0 is the velocity slip factor and K_0 is the thermal slip factor. For $N_0=0=K_0$, one can recover the no- slip case. The stream function ψ is defined by $u=\partial\psi/\partial y$ and $v=-\partial\psi/\partial x$, and therefore, the continuity equation is automatically satisfied. In order to write the governing equations and the boundary conditions in dimensionless form, the following non-dimensional quantities are introduced.

$$\psi = \nu \mathrm{Re}_x^{1/2} f(\xi,\eta),\ \xi = \frac{Gr_x}{\mathrm{Re}_x^2},\ \eta = \frac{y}{x}\mathrm{Re}_x^{1/2},$$

$$\mathrm{Pr} = \frac{\rho \nu c_p}{k},\ Gr = \frac{g\Omega(T_w - T_\infty)x^3}{\nu^2}, \mathrm{Re}_x = \frac{Ux}{\nu}$$

$$\theta(\xi,\eta) = \frac{T - T_\infty}{T_w - T_\infty},\ \beta = \mu_B \frac{\sqrt{2\pi_c}}{p_y} \tag{6}$$

where ξ is the dimensionless stream wise coordinate, $f(\xi,\eta)$ is the dimensionless stream function, $\theta(\xi,\eta)$ is the temperature function, Re_x is the local Reynolds number, Gr_x is the local thermal Grash of parameter, Ω is the coefficient of thermal expansion, ρ is the density of the fluid, $Pr = \frac{\nu}{\alpha}$ is the Prandtl number, T_∞ is the free stream temperature. The local mixed convection parameter ξ is small near the leading edge where the forced convection dominates and large when the buoyancy force dominates the flow field. The stretching velocity of the surface obeys the relation:

$$U(x) = U_0(x) \tag{7}$$

where U_0 is a constant.

In view of Equations (6) and (7), Equations (2)-(4) reduce to the following coupled, nonlinear, dimensionless partial differential equations for momentum and energy for the regime

$$\left(1 + \frac{1}{\beta}\right) f''' + ff'' - f'^2 + \xi\theta = \xi\left(f'\frac{\partial f'}{\partial \xi} - f''\frac{\partial f}{\partial \xi}\right) \tag{8}$$

$$\frac{\theta''}{\mathrm{Pr}} + f\theta' = \xi\left(f'\frac{\partial \theta}{\partial \xi} - \theta'\frac{\partial f}{\partial \xi}\right) \tag{9}$$

The transformed dimensionless boundary conditions are:

$$\text{At } \eta = 0,\ f_w = 0,\ f' = 1 + \left(1 + \frac{1}{\beta}\right) S_f f''(0),\ \theta = 1 + S_T \theta'(0)$$
$$\text{As } \eta \to \infty,\ f' \to 0,\ \theta \to 0 \tag{10}$$

In the above equations, the primes denote the differentiation with respect to η, the dimensionless transverse coordinate, and ξ is the dimensionless tangential coordinate, $S_f = \frac{N_0 \mathrm{Re}_x^{1/2}}{x}$ and $S_T = \frac{K_0 \mathrm{Re}_x^{1/2}}{x}$ are the non-dimensional velocity and thermal slip parameters respectively. Here we assumed the typical values $K_0 = 0.5$, $N_0 = 0.25$ for finding the non-dimensional velocity and thermal slip parameters.

The engineering design quantities of physical interest include the skin-friction coefficient and Nusselt number, which are given by:

$$\frac{1}{2} \mathrm{Re}_x^{1/2} C_f = \left(1 + \frac{1}{\beta}\right) f''(\xi, 0) \tag{11}$$

$$\mathrm{Re}_x^{-1/2} Nu = -\theta'(\xi, 0) \tag{12}$$

3. NUMERICAL SOLUTION

In this study the efficient Keller-Box implicit difference method has been employed to solve the general flow model defined by Equations (8)-(9) with boundary conditions (10). Therefore a more detailed exposition is presented here. This method, originally developed for low speed aerodynamic boundary layers by Keller [29], and has been employed in a diverse range of coupled heat transfer problems. These include Ramachandra Prasad et al. [30], Rao et al. [31] and Beg et al. [32].

Essentially 4 phases are central to the Keller Box Scheme.

These are

a) Reduction of the Nth order partial differential equation system to N first order equations;

b) Finite Difference Discretization;

c) Quasilinearization of Non-Linear Keller Algebraic Equations;

d) Block-tridiagonal Elimination of Linear Keller Algebraic Equations.

Phase a: Reduction of the Nth order partial differential equation system to N first order equations

Equations (8)-(9) subject to the boundary conditions (10) are first written as a system of first-order equations. For this purpose, we reset Equations (8)-(9) as a set of simultaneous equations by introducing the new variables u, v and t:

$$f' = u \tag{13}$$

$$f'' = v \tag{14}$$

$$\theta' = t \tag{15}$$

$$\left(1+\frac{1}{\beta}\right)v' + fv - u^2 + \xi s = \xi\left(u\frac{\partial u}{\partial \xi} - v\frac{\partial f}{\partial \xi}\right) \tag{16}$$

$$\frac{1}{\Pr}t' + ft = \xi\left(u\frac{\partial s}{\partial \xi} - t\frac{\partial f}{\partial \xi}\right) \tag{17}$$

In terms of the dependent variables, the boundary conditions become:

$$\text{At } \eta=0: u=1+\left(1+\frac{1}{\beta}\right)f''(0),\ f=0,\ s=1$$

$$\text{As } \eta\to\infty: u\to 0,\ \ s\to 0 \tag{18}$$

Phase b: Finite difference discretization

A two dimensional computational grid is imposed on the x-η plane as sketched in Figure 2. The stepping process is defined by:

$$\eta_0=0, \eta_j=\eta_{j-1}+h_j, j=1,2,\cdots,J,\ \eta_J=\eta_\infty \tag{19}$$

$$\xi^0=0, \xi^n=\xi^{n-1}+k_n, n=1,2,\cdots,N \tag{20}$$

where k_n and h_j denote the step distances in the ξ and η directions respectively.

If g_j^n denotes the value of any variable at (η_j,ξ^n), then the variables and derivatives of Equations (13)-(17) at $(\eta_{j-1/2},\xi^{n-1/2})$ are replaced by:

$$g_{j-1/2}^{n-1/2}=\frac{1}{4}\left(g_j^n+g_{j-1}^n+g_j^{n-1}+g_{j-1}^{n-1}\right), \tag{21}$$

$$\left(\frac{\partial g}{\partial\eta}\right)_{j-1/2}^{n-1/2}=\frac{1}{2h_j}\left(g_j^n-g_{j-1}^n+g_j^{n-1}-g_{j-1}^{n-1}\right), \tag{22}$$

$$\left(\frac{\partial g}{\partial\xi}\right)_{j-1/2}^{n-1/2}=\frac{1}{2k_n}\left(g_j^n-g_{j-1}^n+g_j^{n-1}-g_{j-1}^{n-1}\right), \tag{23}$$

We now state the finite-difference approximation of Equations (13)-(17) for the mid-point $(\eta_{j-1/2},\xi^n)$, below

$$h_j^{-1}\left(f_j^n-f_{j-1}^n\right)=u_{j-1/2}^n, \tag{24}$$

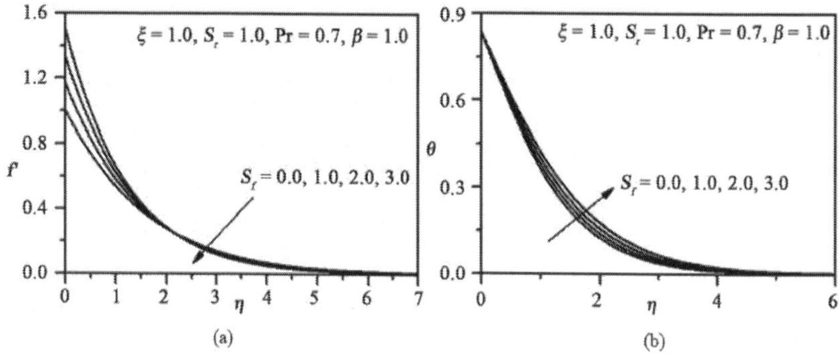

Figure 2. (a) Influence of S_f on the velocity; (b) Influence of S_f on the temperature.

$$h_j^{-1}\left(u_j^n - u_{j-1}^n\right) = v_{j-1/2}^n, \tag{25}$$

$$\left(1+\frac{1}{\beta}\right)\left(v_j - v_{j-1}\right) - \frac{h_j}{4}(1+\alpha)\left(u_j + u_{j-1}\right)^2 + \frac{\alpha h_j}{2} v_{j-1/2}^{n-1}\left(f_j + f_{j-1}\right)$$
$$+\frac{(1+\alpha)h_j}{4}\left[\left(f_j + f_{j-1}\right)\left(v_j + v_{j-1}\right)\right] - \frac{\alpha h_j}{2} f_{j-1/2}^{n-1}\left(v_j + v_{j-1}\right) + \frac{\xi h_j}{2}\left(s_j + s_{j-1}\right) = \left[R_1\right]_{j-1/2}^{n-1} \tag{26}$$

$$h_j^{-1}\left(\theta_j^n - \theta_{j-1}^n\right) = t_{j-1/2}^n, \tag{27}$$

$$\frac{1}{\Pr}\left(t_j - t_{j-1}\right) + \frac{(1+\alpha)h_j}{4}\left[\left(f_j + f_{j-1}\right)\left(t_j + t_{j-1}\right)\right] - \frac{\alpha h_j}{4}\left[\left(u_j + u_{j-1}\right)\left(s_j + s_{j-1}\right)\right]$$
$$+\frac{\alpha h_j}{2} s_{j-1/2}^{n-1}\left(u_j + u_{j-1}\right) - \frac{\alpha h_j}{2} u_{j-1/2}^{n-1}\left(s_j + s_{j-1}\right) - \frac{\alpha h_j}{2} f_{j-1/2}^{n-1}\left(t_j + t_{j-1}\right)$$
$$+\frac{\alpha h_j}{2} t_{j-1/2}^{n-1}\left(f_j + f_{j-1}\right) = \left[R_2\right]_{j-1/2}^{n-1} \tag{28}$$

where we have used the abbreviations

$$\alpha = \frac{\xi^{n-1/2}}{k_n} \tag{29}$$

$$[R_1]^{n-1}_{j-1/2} = -h_j\left[\left(1+\frac{1}{\beta}\right)\left(\frac{v_j - v_{j-1}}{h_j}\right) + (1-\alpha)\left(f_{j-1/2}v_{j-1/2}\right) - (1-\alpha)\left(u_{j-1/2}\right)^2 + \xi\left(s_{j-1/2}\right)\right] \tag{30}$$

$$[R_2]^{n-1}_{j-1/2} = -h_j\left[\frac{1}{\Pr}\left(\frac{t_j - t_{j-1}}{h_j}\right) + (1-\alpha)\left(f_{j-1/2}t_{j-1/2}\right) + \alpha\left(u_{j-1/2}s_{j-1/2}\right)\right] \tag{31}$$

The boundary conditions are

$$f_0^n = u_0^n = 0,\ \theta_0^n = 1,\ u_J^n = 0,\ \theta_J^n = 0 \tag{32}$$

Phase c: Quasilinearization of non-linear Keller algebraic equations

If we assume $f_j^{n-1}, u_j^{n-1}, v_j^{n-1}, s_j^{n-1}, t_j^{n-1}$ to be known for $0 \le j \le J$, Equations (24)-(28) are a system of 5J + 5 equations for the solution of 5J + 5 unknowns $f_j^n, u_j^n, v_j^n, s_j^n, t_j^n$, $j = 0,1,2,\cdots,J$. This non-linear system of algebraic equations is linearized by means of Newton's method as explained in Keller [29] and Prasad et al. [33].

Phase d: Block-tridiagonal elimination of linear Keller algebraic equations

The linear system (24)-(28) can now be solved by the block-elimination method, since they possess a block- tridiagonal structure. Commonly, the block-tridiagonal structure consists of variables or constants, but here, an interesting feature can be observed, namely that it consists of block matrices. The complete linearized system is formulated as a block matrix system, where each element in the coefficient matrix is a matrix itself. Then, this system is solved using the efficient Keller-box method. The numerical results are affected by the number of mesh points in both directions. After some trials in the η-direction (radial coordinate) a larger number of mesh points are selected whereas in the ξ direction (tangential coordinate) significantly less mesh points are utilized. η_{max} has been set at 10 and this defines an adequately large value at which the prescribed boundary conditions are satisfied. ξ_{max} is set at 3.0 for this flow domain. Mesh independence is therefore achieved in the present com- putations. The computer program of the algorithm is executed in MATLAB running on a PC. The method demonstrates excellent stability, convergence and consistency, as elaborated by Keller [29] and this system is developed by Cebeci and Bradshaw [34].

4. RESULTS AND DISCUSSIONS

Comprehensive solutions have been obtained and are presented in Figures 2-7. The numerical problem comprises 2 independent variables (x, h), 2 dependent fluid dynamic variables (f,θ) and 5 thermo physical and body force control parameters $\Pr, S_f, S_T, \beta, \xi$. In the present computations, the following default parameters are prescribed (unless otherwise stated): Pr = 0.71, S_f = 0.5, S_T = 1.0, b = 1.0, x = 1.0.

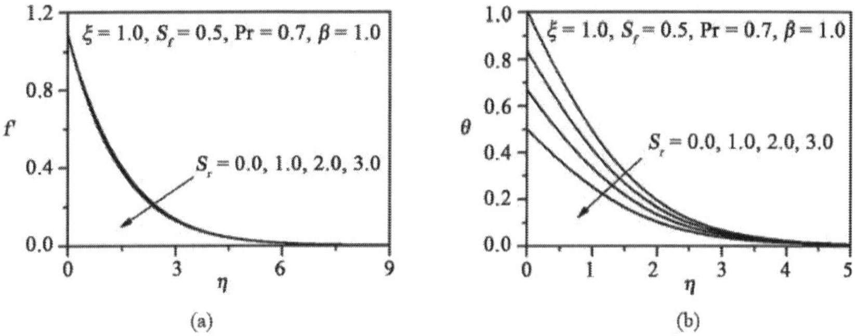

Figure 3. (a) Influence of S_T on the velocity; (b) Influence of S_T on the temperature.

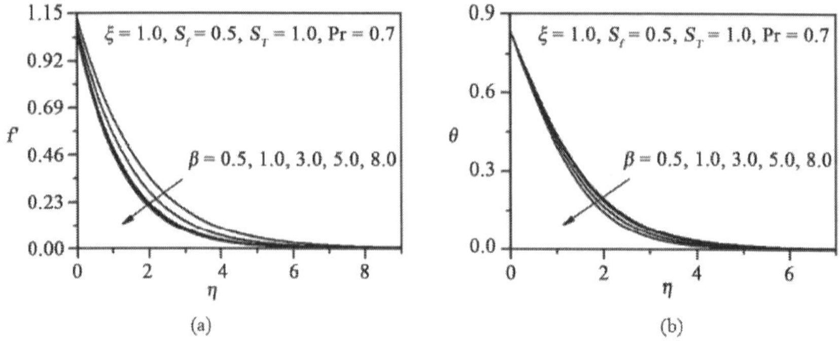

Figure 4. (a) Influence of β on the velocity; (b) Influence of β on the temperature.

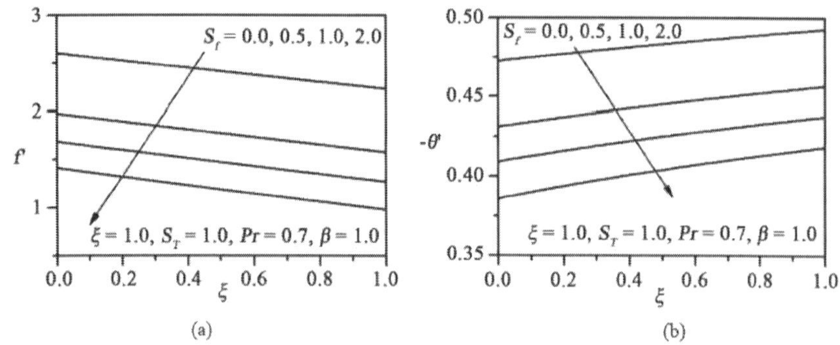

Figure 5. (a) Effect of S_f on the Skin-friction coefficient results; (b) Effect of S_f on the Nusselt number results.

In Figure 2(a) and Figure 2(b), the influence of velocity slip parameter on velocity and temperature is illustrated. In Figure 2(a) the dimensionless velocity component at the wall reduces with increase in slip parameter and hence there will be a decrease in the boundary layer thickness. The velocity profiles damped out a bit slower for the high amount of slip parameters, because of an interception which exhibits among them. Figure 2(b) indicates that an increase in slip parameter tends to increase temperature in the flow field. By increasing S_f, thermal boundary layer thickness enhances.

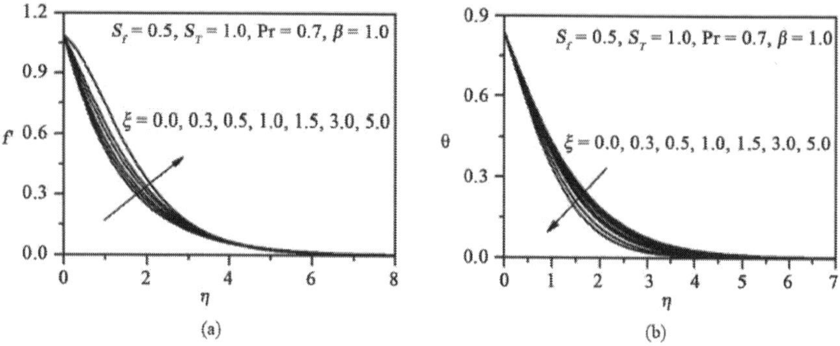

Figure 6. (a) Influence of ξ on the velocity; (b) Influence of ξ on the temperature.

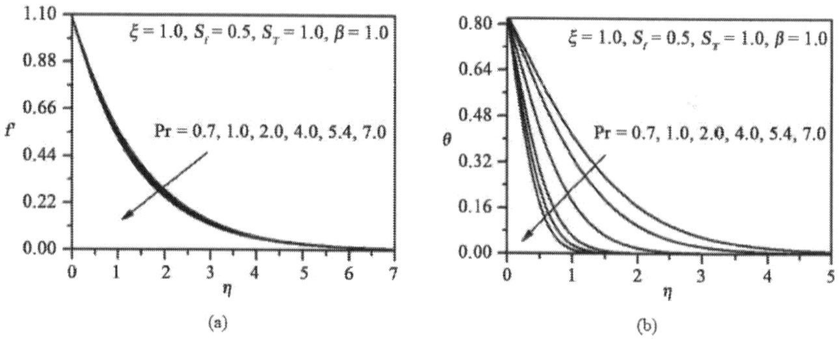

Figure 7. (a) Influence of Pr on the velocity; (b) Influence of Pr on the temperature.

The variation of velocity and temperature with the transverse coordinate (h), over the thermal slip parameter S_T is illustrated in Figure 3(a) and Figure 3(b). The response of velocity is much more consistent than for the case of changing velocity slip parameter, it is strongly decreased for all locations in the radial direction. The peak velocity accompanies the case of no thermal slip ($S_T = 0$). The maximum deceleration corresponds to the case of strongest thermal slip ($S_T = 3$). Temperatures (Figure 3(b)) are also strongly depressed with increasing thermal slip. The maximum effect is observed at the wall. Further into the free stream, all temperature profiles converge smoothly to the vanishing value. Figure 4(a) and Figure 4(b) depict the effect of the Casson fluid parameter β on velocity and temperature. Actually, with an increase in non-Newtonian Casson parameter (β), it produces resistance in the fluid flow. An increase in β implies a decrease in yield stress P_y of the Casson fluid and increase in the value of plastic dynamic viscosity μ_B; this effect creates resistance in the flow of fluid. It is further noted that velocity decreases as the Casson fluid parameter β increases. In Figure 4(b) it is shown that the effect of β causes decreases in temperature. The effect of velocity slip parameter S_f on stretching surface shear stress (f''), local Nusselt number variation $(-\theta')$ are presented in Figure 5(a) and Figure 5(b). In consistency with the earlier graphs described for velocity evolution, with an increase in S_f, wall shear stress is consistently reduced i.e. the flow is decelerated along the stretching surface. The impact of wall slip is therefore significant on the boundary layer characteristics of Casson flow from a surface. With an increasing S_f, the local Nusselt number is also considerably decreased and profiles are generally monotonic decays. Maximum local Nusselt number always arises at the stretching surface and is minimized with proximity to the greater distance from the stretching surface. In both Figure 5(a) and Figure 5(b), skin friction coefficient and local Nusselt number are maximized for the case of no-slip i.e. $S_f = 0$. In Figure 6(a) and Figure 6(b), the variation of velocity and temperature fields with different ξ values is shown. Close to the stretching surface, velocity (f') is found to be maximized closer to the stretching surface and minimized with progressive distance away from it i.e. the flow is decelerated with increasing ξ. However further from the wall, a marked acceleration in the flow is generated with greater distance from the surface i.e. velocity values are higher for higher values of ξ. Temperature q is found to noticeably decrease through the boundary layer with increasing ξ values; as such the fluid regime is cooled most efficiently at the stretching surface and heated increasingly as we progress around the stretching surface periphery upwards. The effect of Prandtl number (Pr) on the primitive flow variables of velocity and temperature is shown inFigure 7(a) and Figure 7(b). Prandtl number signifies the ratio of viscous diffusion to thermal diffusion in the boundary layer regime. With greater Pr values, viscous diffusion rate exceeds thermal diffusion rate. An increase in Pr from 0.7 through 1.0, 2.0, 4.0, 5.4 to 7.0, strongly depresses velocities (Figure 7(a)) in the regime. For Pr < 1, thermal diffusivity exceeds momentum diffusivity i.e. heat will diffuse faster than momentum. For Pr = 1.0, both the viscous and energy diffusion rates will be the same as will the thermal and velocity boundary layer thicknesses. With increasing Pr values, temperature as shown in Figure 7(b), is markedly reduced throughout the boundary layer.

To validate the present solutions, we compare the present model with the earlier Newtonian model of Merkin [35] and we observe that an excellent agreement between the previous results as shown in Table 1.

5. CONCLUSIONS

In this study, numerical solutions have been presented for flow and heat transfer of Casson fluid from a permeable isothermal stretching surface with partial slip. The model has been developed to simulate food stuff transport processes in industrial manufacturing operations. A robust, extensively-validated and implicit finite difference numerical scheme has been implemented to solve the transformed and dimensionless velocity and thermal boundary layer equations, subject to physically realistic boundary conditions. The results in summary have shown that, when increasing the velocity slip parameter, velocity, skin friction and Nusselt number decrease, but the temperature increases. A significant finding of this study is that flow separation can be controlled by increasing the value of Casson fluid parameter as well as by increasing Prandtl number.

Table 1. Values of the local heat transfer coefficient for various values of ξ with $\beta \rightarrow \infty$, $f_w = N_0 = K_0 = 0$.

ξ	$-\theta'(\xi,0)$	
	Merkin [35]	Present results
0.0	0.4212	0.4198
0.2	0.4204	0.4201
0.4	0.4182	0.4191
0.6	0.4145	0.4149
0.8	0.4093	0.4098
1.0	0.4025	0.4034
1.2	0.3942	0.3945
1.4	0.3843	0.3847
1.6	0.3727	0.3729
1.8	0.3594	0.3598
2.0	0.3443	0.3448
2.2	0.3270	0.3272
2.4	0.3073	0.3079
2.6	0.2847	0.2852
2.8	0.2581	0.2588
3.0	0.2252	0.2254
π	0.1963	0.1964

The current study has been confined to steady-state flow i.e. ignored transient effects and neglected thermal radiation heat transfer effects [36]. Generally, very stable and accurate solutions are obtained with the present finite difference code and it is envisaged that other non-Newtonian flows will be studied using this methodology in the future, including Maxwell upper convected fluids [37], and couple stress fluids [38]. These aspects are also of relevance to rheological food processing simulations and will be considered in future investigations.

ACKNOWLEDGEMENTS

The authors are grateful to the reviewers for giving their constructive comments for improving this article. The work is supported by the University Grants Commission-SERO. The authors are thankful to UGC-NEWDELHI, S.V University, Tirupati and management of MITS, Madanapalle.

REFERENCES

1. Sakiadis, B.C. (1961) Boundary-Layer Behavior on Continuous Solid Surfaces: I. Boundary-Layer Equations for Two-Dimensional and Axisymmetric Flow. AIChE Journal, 7, 26-28.

2. Sakiadis, B.C. (1961) Boundary-Layer Behavior on Continuous Solid Surfaces: II. Boundary-Layer Equations on a Continuous Flat Surface. AIChE Journal, 7, 221-225.

3. Erickson. L.E., Fan. L.T. and Fox, V.G. (1966) Heat and Mass Transfer on Moving Continuous Flat Plate with Suction or Injection. Industrial Engineering Chemistry Fundamentals, 5, 19-25.

4. Gireesha, B.J., Roopa, G.S. and Bagewadi, C.S. (2011) Boundary Layer Flow of an Unsteady Dusty Fluid and Heat Transfer over a Stretching Sheet with Non-Uniform Heat Source/Sink. Scientific Research, 3, 726-735.

5. Chen, T.S. and Strobel, F.A. (1980) Buoyancy Effects in Boundary Layer Adjacent to a Continuous, Moving Horizontal Flat Plate. Journal of Heat Transfer, 102, 170-172.

6. Grubka, L.J. and Bobba, K.M. (1985) Heat Transfer Characteristics of a Continuous, Stretching Surface with Variable Temperature. ASME J. Heat Transfer, 107, 248-250.

7. Dutta, B.K., Roy, P. and Gupta, A.S. (1985) Temperature Field in Flow over a Stretching Surface with Uniform Heat Flux. International Communications in Heat and Mass Transfer, 12, 89-94.

8. Chen, C.K. and Char, M.I. (1988) Heat Transfer of a Continuous Stretching Surface with Suction or Blowing. Journal of Mathematical Analysis and Applications, 135, 568-580.

9. Karwe, M.V. and Jaluria, Y. (1988) Fluid Flow and Mixed Convection Transport from a Moving Plate in Rolling and Extrusion Processes. ASME J. Heat Transfer, 110, 655-661.

10. Karwe, M.V. and Jaluria, Y. (1991) Numerical Simulation of Thermal Transport Associated With a Continuously Moving Flat Sheet in Materials Processing. ASME J. Heat Transfer, 113, 612-619.

11. Patil, P.M., Roy, S. and Pop, I. (2010) Unsteady Mixed Convection Flow over a Vertical Stretching Sheet in a Parallel Free Stream with Variable Wall Temperature. International Journal of Heat and Mass Transfer, 53, 4741-4748.

12. Rajeswari, V., Kumari, M. and Nath, G. (1993) Unsteady Three-Dimensional Boundary Layer Flow Due to a Stretching Surface. Actamechanica, 98, 123-141.

13. Ali, M. and Al-Yousef, F. (2002) Laminar Mixed Convection Boundary Layers Induced by a Linearly Stretching Permeable Surface. International Journal of Heat and Mass Transfer, 45, 4241-4250.

14. Partha, M.K., Murthy, P.V.S.N. and Rajasekhar, G.P. (2005) Effect of Viscous Dissipation on the Mixed Convection Heat Transfer from an Exponentially Stretching Surface. Heat and Mass Transfer, 41, 360-366.

15. Schowalter, W.R. (1978) Mechanics of Non-Newtonian Fluids. Pergamon Press, New York.

16. Rana, P. and Bhargava, R. (2012) Flow and Heat Transfer of a Nanofluid over a Nonlinearly Stretching Sheet: A Numerical Study. Communications in Nonlinear Science and Numerical Simulation, 17, 212-226.

17. Nazar, M., Fetecau, C., Vieru, D. and Fetecau, C. (2010) New Exact Solutions Corresponding to the Second Problem of Stokes for Second Grade Fluids. Nonlinear Analysis: Real World Applications, 11, 584-591.

18. Fetecau, C., Hayat, T., Zierep, J. and Sajid, M. (2011) Energetic Balance for the Rayleigh—Stokes problem of an Oldroyd-B fluid. Nonlinear Analysis: Real World Applications, 12, 1-13.

19. Wang, S.W. and Tan, W.C. (2008) Stability Analysis of Double-Diffusive Convection of Maxwell Fluid in a Porous Medium Heated from Below. Physics Letters A, 372, 3046-3050.

20. Tan, W.C. and Xu, M.Y. (2004) Unsteady Flows of a Generalized Second Grade Fluid with the Fractional Derivative Model between Two Parallel Plates. Acta Mechanica Sinica, 20, 471-476.

21. Zhang, Z.Y., Fu, C.J., Tan, W.C. and Wang, C.Y. (2007) On Set of Oscillatory Convection in a Porous Cylinder Saturated with a Viscoelastic Fluid. Physics of Fluids, 19, 98-104.

22. Rashidi, M.M., Chamkha, A.J. and Keimanesh, M. (2011) Application of Multi-Step Differential Transform Method on Flow of a Second Grade Fluid over a Stretching or Shrinking Sheet. American Journal of Computational Mathematics, 6, 119-128.

23. Ali, N., Hayat, T. and Asghar, S. (2009) Peristaltic Flow of Maxwell Fluid in a Channel with Compliant Walls. Chaos, Solitons & Fractals, 39, 407-416.

24. Attia, H.A. and Seddeek, M.A. (2007) On the Effectiveness of Uniform Suction or Injection on Two Dimensional Stagnation-Point Flow towards a Stretching Surface with Heat Generation. Chemical Engineering Communications, 194, 553-564.

25. Hussain, M., Hayat, T., Asghar, S. and Fetecau, C. (2010) Oscillatory Flows of Second Grade Fluid in a Porous Space. Nonlinear Analysis: Real World Applications, 11, 2403-2414.

26. Casson, N. (1959) In Reheology of Dipersed System. Peragamon Press, Oxford.

27. Nakamura, M. and Sawada, T. (1988) Numerical Study on the Flow of a Non-Newtonian Fluid through an Axisymmetric Stenosis. Journal of Biomechanical Engineering, 110, 137-143.

28. Samir Kumar, N. (2013) Analytical Solution of MHD Stagnation-Point Flow and Heat Transfer of Casson Fluid over a Stretching Sheet with Partial Slip. ISRN Thermodynamics, 2013, Article ID: 108264.

29. Keller, H.B. (1970) A New Difference Method for Parabolic Problems. In: Bramble, J., Ed., Numerical Methods for Partial Differential Equations, Academic Press, New York, 327-350.

30. Prasd, V.R., Vasu, B. and Beg, O.A. (2011) Thermo-Diffusion and Diffusion-Thermo Effects on Boundary Layer Flows. LAP Lambert Academic Publishing GmbH & Co. KG, Saarbrücken.

31. Rao, A.S., Prasad, V.R., Reddy, N.B. and Bég, O.A. (2013) Heat Transfer in a Casson Rheological Fluid from a Semi-infinite Vertical Plate with Partial Slip. Heat Transfer-Asian Research, 44, 272-291.

32. Bég, O.A., Prasad, V.R., Vasu, B., Reddy, N.B., Li, Q. and Bhargava, R. (2011) Free Convection Heat and Mass Transfer from an Isothermal Sphere to a Micropolar Regime with Soret/Dufour

Effects. International Journal of Heat and Mass Transfer, 54, 9-18.

33. Prasad, V.R., Rao, A.S., Reddy, N.B., Vasu, B. and Beg, O.A. (2013) Modelling Laminar Transport Phenomena in a Casson Rheological Fluid from a Horizontal Circular Cylinder with Partial Slip. Proceedings of the Institution of Mechanical Engineers, Part E: Journal of Process Mechanical Engineering, 227, 309-326.

34. Cebeci, T. and Bradshaw, P. (1984) Physical and Computational Aspects of Convective Heat Transfer. Springer, New York.

35. Merkin, J.H. (1977) Free Convection Boundary Layers on Cylinders of Elliptic Cross Section. Journal of Heat Transfer, 99, 453-457.

36. Prasad, V.R., Vasu, B., Prashad, D.R. and Bég, O.A. (2012) Thermal Radiation Effects on Magneto-Hydrodynamic Heat and Mass Transfer from a Horizontal Cylinder in a Variable Porosity Regime. Journal of Porous Media, 15, 261- 281.

37. B´eg, O.A. and Makinde, O.D. (2011) Viscoelastic Flow and Species Transfer in a Darcian High-Permeability Channel. Journal of Petroleum Science and Engineering, 76, 93-99.

38. Kairi, R.R. and Murthy, P.V.S.N. (2012) Effect of Melting on Mixed Convection Heat and Mass Transfer in a Non-Newtonian Fluid Saturated Non-Darcy Porous Medium. Journal of Heat Transfer, 134, Article ID: 042601.

9

ENTROPY ASSESSMENT ON DIRECT CONTACT CONDENSATION OF SUBSONIC STEAM JETS IN A WATER TANK THROUGH NUMERICAL INVESTIGATION

Yu Ji [1,2], Hao-Chun Zhang [1,*], Jian-Fei Tong [3], Xu-Wei Wang [4], Han Wang [4] and Yi-Ning Zhang [1]

[1]School of Energy Science and Engineering, Harbin Institute of Technology, Harbin 150001, China

[2]Institute of Nuclear and New Energy Technology, Tsinghua University, Beijing 100084, China

[3]Institute of High Energy Physics, Chinese Academy of Sciences, Beijing 100049, China

[4]State Nuclear Power Technology R & D Center, Beijing 102209, China

ABSTRACT

The present article analyzes the dissipation characteristics of the direct contact condensation (DCC) phenomenon that occurs when steam is injected into a water tank at a subsonic speed using a new modeling approach for the entropy generation over the calculation domain. The developed entropy assessment model is based on the local equilibrium hypothesis of non-equilibrium thermodynamics. The fluid flow and heat transfer processes are investigated numerically. To describe the condensation and evaporation process at the vapor-liquid interface, a phase change model originated from the kinetic theory of gas is implemented with the mixture model for multiphase flow in the computational fluid dynamics (CFD) code ANSYS-FLUENT. The CFD predictions agree well with the published works, which indicates the phase change model combined with the mixture model is a promising way to simulate the DCC phenomenon. In addition, three clear stages as initial stage, developing stage and oscillatory stage are discriminated from both the thermal-hydraulic results and the entropy generation information. During different stages, different proportion of the entropy generation rate owing to heat transfer, viscous direct dissipation, turbulent dissipation and inner phase change in total entropy generation rate is estimated, which is favorable to deeper understanding the irreversibility of DCC phenomenon, designing and optimizing the equipment involved in the process.

KEYWORDS

entropy generation; subsonic steam jet; direct contact condensation; transport process; numerical investigation

1. INTRODUCTION

Direct contact condensation (DCC) phenomena exists in many industrial facilities, such as the suppression pools of boiling water reactors, the emergency core cooling systems (ECCSs) for nuclear reactors, direct contact heat exchangers, and thus have attracted great interest and received much attention in the past several decades. Generally, steam is discharged into subcooled water at high speed, then contacts with water and condenses at the interface directly in a DCC process. Knowledge including the jet shape, plume length, temperature and pressure fluctuation is primarily required for engineering design, and a reliable prediction of the process is essential, because the rapid condensation could induce accidents under certain conditions during the DCC process.

A wide variety of theoretical and experimental studies have been performed to investigate the DCC phenomenon, in attempting to provide technical support for equipment design and operation. Kerney *et al.* [1] and Weimer *et al.* [2] investigated the submerged sonic steam jet in quiescent sub-cooled water tanks by experimental and theoretical approaches, and a semi-empirical correlation to calculate the steam penetration length was developed. Aya and Nariai [3] proposed the correlation of heat transfer coefficients under low mass flux conditions through the experimental investigation of steam jets injected into cold water. Chan and Lee [4] proposed the regime maps at low steam mass flux, and the condensation mode was categorized into several classes. On top of the experimental investigations, Computational Fluid Dynamics (CFD) technique is also a powerful tool to analyze two phase flow processes. With the rapid development of CFD, several numerical studies have been performed to investigate the DCC phenomenon. Gulawani *et al.* [5,6] analyzed the temperature profile, flow pattern and heat transfer characteristics by using a thermal phase change model of commercial CFD code under a three dimensional simulation of DCC process, and the corresponding experiment was carried out to validate the results by CFD prediction. Dahikar *et al.* [7] investigated the flow pattern and temperature distribution of DCC using PIV (Particle Image Velocimetry), PLIF (Planar Laser Induced Fluorescence) and CFD, and during the simulation, two turbulence models,*i.e.*, k-ε and Large Eddy Simulation (LES) were employed, through which the flow field, temperature field and phase change information can be monitored instantaneously. Shah *et al.* [8] simulated the DCC of supersonic steam jet under the Euler-Euler framework of FLUENT 6.3, and the results agreed fairly well with the published experimental data. Recently, Li *et al.*[9] performed the simulation using the Volume of Fluid (VOF) model for multiphase flow and LES turbulent model in FLUENT, through which, good consistance with the available experiment data was obtained.

As an effective method to estimate transport phenomena, the thermodynamic approach has been widely applied to viscous flow, heat and mass transfer processes [10,11,12,13,14]. The traditional way in two phase flow emphasizes the essential thermal-hydraulic parameters, while the second law of thermodynamics provides another powerful tool to investigate phase changes [15,16,17]. The condensation process is accompanied by thermodynamic irreversibility resulting from fluid friction, heat transfer over finite temperature gradients and inner phase change in fluid flow domain; therefore, a relationship exists between entropy generation and the energy devaluation in the process [18,19]. Naterer and Adeyinka [20] proposed a correlation for entropy generation and exergy in film condensation; and in this way, entropy generation can be used as an important parameter in optimizing a two-phase system. Thiel [21] examined the mechanism of entropy generation in a condenser with high fractions of non-condensable gases using scaling and boundary layer technique, aiming at providing a criterion for minimum entropy generation, which is valuable in engineering analysis. Revellin *et al.* [22] applied the local entropy generation rate formulation catering to the saturated two phase flow to analyze the performance of heat exchangers in the various conditions. Furthermore, Herwig [23] and Wenterodt [24] introduced the concept of entropic potential of energy, which consider the energy transfer within a certain process from a broader perspective.

From the literature review above, it should be highlighted that, although numerous investigations for DCC have been performed, essentially, it is still not well understood. In the current work, a transient simulation for DCC process of steam is performed through the mixture model in the commercial CFD code, ANSYS-FLUENT, and a phase change model derived from the kinetic theory of gas is implemented to the solver through User Defined Functions (UDF).

Moreover, the entropy generation analysis approach is used to estimate the irreversibility of DCC, which is expected to be a powerful tool for designing and optimizing the equipment involved in the phenomenon.

2. GEOMETRY MODEL

DCC processes appear in various types of facilities, but it can be generally considered as a fundamental transport processes in a specified geometry for the same mechanism. In this work, the simulation is performed on a simplified hexagonal tank of 560 mm × 560 mm × 870 mm dimensions. The tank is full of subcooled water, while saturated steam is ejected from a vertical downward pipe with a diameter of 5 mm. The pipe is centrally located in the top boundary of water tank and it is submerged vertically 250 mm in the water. A schematic view of the geometry model is shown in Figure 1.

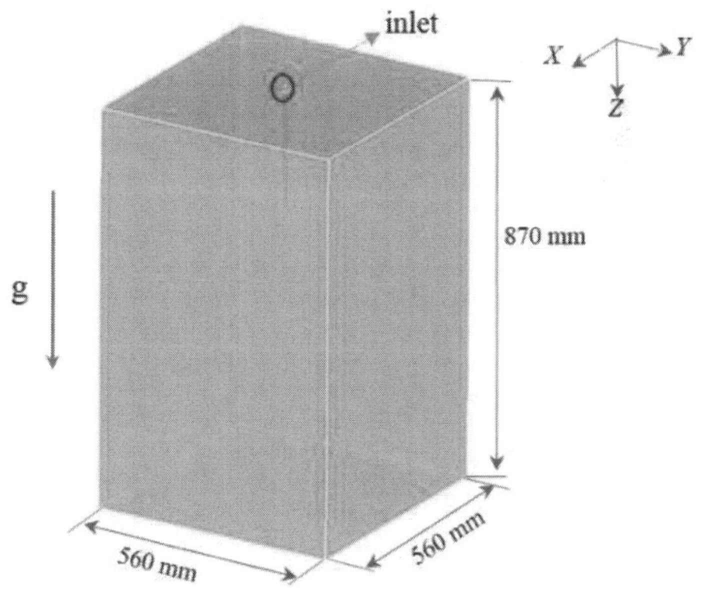

Figure 1. Schematic view of physical model used in the simulation of DCC process.

3. MATHEMATICAL MODEL

In the current work, the mixture model concerning the slip velocity is employed, through which the intensive shear interaction can be simulated in the non-homogenous two phase flow. Liquid water is treated as the primary phase, while the vapor is secondary phase. In addition, both phases are regarded as inter-penetrating continua in the simulation. The condensation and evaporation process occurring at the interface is described by the phase change model based on the kinetic theory of gas. For the turbulence model, the standard *k*-*ε* model with standard wall functions treatment is adopted. Moreover, an entropy generation assessment approach is proposed to analyze the irreversibility during the DCC process.

3.1. Mixture Model

The mixture model [25] is used to simulate separated phase flow, but it assumes local equilibrium over short spatial length scales. Compared to the full Eulerian model in [6,7,8], the model solves smaller number of variables; and therefore, it is assumed to be a good substitute for Eulerian model in several cases, including the simulation in the present work.

3.1.1. Continuity Equation

The continuity equation for mixture describing the mass flux into and out of a control volume boundary as well as the internal viration of mass is:

$$\frac{\partial \rho_m}{\partial t} + \nabla \cdot (\rho_m v_m) = 0 \tag{1}$$

where ρ_m and $\boldsymbol{v}_m$ are the density and mass-averaged velocity for mixture, which are defined respectively as follows:

$$\rho_m = \sum_{q=1}^{2} \alpha_q \rho_q, v_m = \frac{\sum_{q=1}^{2} \alpha_q \rho_q v_q}{\rho_m} \tag{2}$$

where α_q is the volume fraction of phase q.

3.1.2. Momentum Equation

The momentum equation for mixture solved throughout the calculating domain is given by:

$$\frac{\partial}{\partial t}(\rho_m v_m) + \nabla \cdot (\rho_m v_m v_m) = -\nabla p + \nabla \cdot \left(\mu_m \left(\nabla v_m + \nabla v_m^T\right)\right) + \rho_m g + F_{dr} \tag{3}$$

The terms on the left side of equation are the inner variation of momentum with time and convective momentum flux crossing the boundaries of control volumes, respectively. The terms on the right side are overall pressure, stresses, gravitational force, and interaction forces between phases due to slip velocity, in proper order. The interaction force is defined as follows:

$$F_{dr} = \nabla \cdot \left(\sum_{q=1}^{2} \alpha_q \rho_q v_{dr,q} v_{dq,q}\right) = \nabla \cdot \left(\sum_{q=1}^{2} \alpha_q \rho_q (v_q - v_m)(v_q - v_m)\right) \tag{4}$$

where μ_m is the viscosity of mixture, defined as:

$$\mu_m = \sum_{q=1}^{2} \alpha_q \mu_q \tag{5}$$

3.1.3. Energy Equation

The energy equation for the mixture is:

$$\frac{\partial}{\partial t}\sum_{q=1}^{2}\left(\alpha_q\rho_q E_q\right)+\nabla\cdot\sum_{q=1}^{2}\left(\alpha_q v_q\left(\rho_q E_q\right)\right)=\frac{\partial}{\partial t}p+\nabla\cdot\left(\kappa_{eff}\nabla T\right)+\sum_{q=1}^{2}S_{E_q} \quad (6)$$

where E_q is the total energy of phase q; κ_{eff} is the effective thermal conductivity, which can be denoted as:

$$\kappa_{eff}=\sum_{q=1}^{2}\alpha_q\left(\kappa_q+\kappa_t\right) \quad (7)$$

in which, κ_q is the thermal conductivity of phase q; κ_t is the turbulent thermal conductivity, determined by the turbulence closure model; S_{Eq} is the volumetric heat source for phase q.

3.2. Turbulence Model

The standard k-ε model [25] is chosen for turbulence model. In the formulation, the conservation equations for turbulent kinetic energy k and the turbulent dissipation ε are expressed as:

$$\frac{\partial}{\partial t}\left(\rho_m k\right)+\nabla\cdot\left(\rho_m v_m k\right)=\nabla\cdot\left(\frac{\mu_{t,m}}{\sigma_k}\nabla k\right)+G_{k,m}-\rho_m\varepsilon \quad (8)$$

$$\frac{\partial}{\partial t}\left(\rho_m\varepsilon\right)+\nabla\cdot\left(\rho_m v_m\varepsilon\right)=\nabla\cdot\left(\frac{\mu_{t,m}}{\sigma_\varepsilon}\nabla\varepsilon\right)+\frac{\varepsilon}{k}\left(C_{1\varepsilon}G_{k,m}-C_{2\varepsilon}\rho_m\varepsilon\right) \quad (9)$$

where $\mu_{t,m}$ is the turbulent viscosity of mixture, defined as:

$$\mu_{t,m}=C_\mu\rho_m\frac{k^2}{\varepsilon} \quad (10)$$

$G_{k,m}$ is the production of turbulent kinetic energy, and it is computed from:

$$G_{k,m} = \mu_{t,m}\left(\nabla v_m + \nabla v_m^T\right) : \nabla v_m \tag{11}$$

The value for the constant above are:

$\sigma_k = 1.0$, $\sigma_\varepsilon = 1.3$, $C_\mu = 0.09$, $C_{1\varepsilon}=1.44$, $C_{2\varepsilon}=1.92$.

3.3. Phase Change Model

According to the kinetic theory of gas, the molecules emitted from both vapor and water will be absorbed totally when they move to the equilibrium interface, and the mass transfer rate between the phases can be calculated in Hertz-Knudsen relation; however, the interface is usually in non-equilibrium state, and the reflection exists there, as is shown in Figure 2.

Then the simplified Hertz-Knudsen-Schrage Equation gives the following form for a smooth phase interface [26,27]:

$$\begin{cases} j^+ = \gamma_c\sqrt{\dfrac{M}{2\pi RT_{sat}}}\left(p_g - p_{sat}\right) \\ j^- = \gamma_e\sqrt{\dfrac{M}{2\pi RT_{sat}}}\left(p_l - p_{sat}\right) \end{cases} \tag{12}$$

In Equation (12), the evaporation coefficient and condensation coefficient are defined respectively by Knudsen and Prüger [28,29] as:

$$\gamma_e = \frac{\text{number of molecules transferred to the vapor phase}}{\text{number of molecules emitted from the liquid phase}}$$

$$\gamma_c = \frac{\text{number of molecules absorbed by the liquid phase}}{\text{number of molecules impinging on the liquid phase}}$$

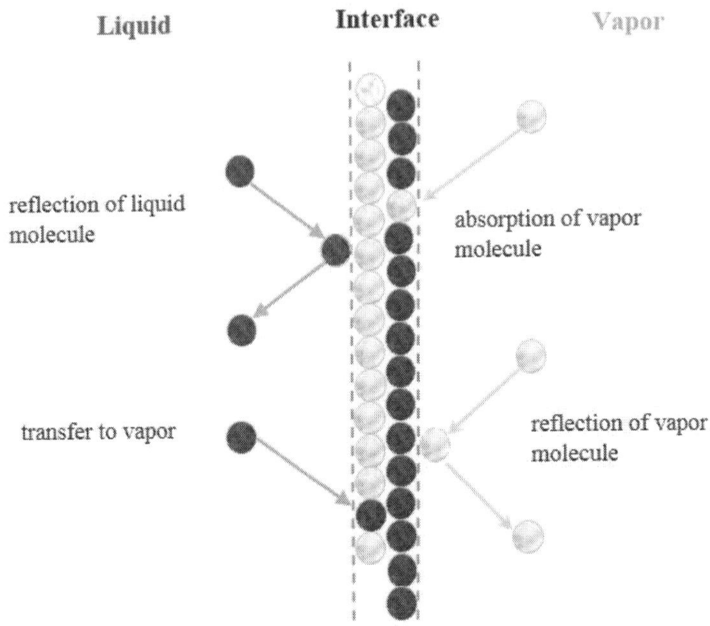

Figure 2. Molecular mechanisms of condensation and evaporation at vapor-liquid interface.

The pressure and temperature can be correlated in the vicinity of the saturation condition using Clapeyron-Clausius equation, which is defined as:

$$\frac{\mathrm{d}p}{\mathrm{d}T}=\frac{L}{T\left(v_g-v_l\right)} \tag{13}$$

in which, L is the latent heat. In this view, the condensation and evaporation rate can be rewritten as:

$$\begin{cases} j^+=\gamma_c\sqrt{\dfrac{M}{2\pi RT_{sat}\left(p_g\right)}}\dfrac{\rho_l\rho_g}{\rho_l-\rho_g}L\dfrac{T_{sat}\left(p_g\right)-T}{T_{sat}\left(p_g\right)} \\ j^-=\gamma_e\sqrt{\dfrac{M}{2\pi RT_{sat}\left(p_l\right)}}\dfrac{\rho_l\rho_g}{\rho_l-\rho_g}L\dfrac{T_{sat}\left(p_l\right)-T}{T_{sat}\left(p_l\right)} \end{cases} \tag{14}$$

where T is the real temperature, T_{sat} (p_g) is the saturated temperature corresponding to the pressure p_g, and T_{sat} (p_l) is the saturated temperature corresponding to the pressure p_l.

It is also assumed that vapor bubble is in dispersed phase when it condenses, and liquid droplet is in dispersed phase when it evaporates at the interface. Furthermore, the dispersed phases contain spherical particles of the same diameter, and then the volumetric interfacial area can be derived as the following form:

$$A_{fg} = \frac{\alpha_p A_p}{V_{cell}} = \alpha_p \frac{\pi d_p^2}{6\pi d_p^3} = 6\frac{\alpha_p}{d_p} \tag{15}$$

in which, d_p is the diameter of dispersed phase, and α_p is volume fraction of dispersed phase. Therefore, the volumetric condensation rate and evaporation rate are given by:

$$\begin{cases} J^+ = j^+ A_{fg} = \underbrace{\frac{6\gamma_c}{d_g}\sqrt{\frac{M}{2\pi R T_{sat}(p_g)}} \frac{\rho_l}{\rho_l - \rho_g}}_{\Re_c} L\alpha_g \rho_g \frac{T_{sat}(p_g) - T}{T_{sat}(p_g)} \\ J^- = j^- A_{fg} = \underbrace{\frac{6\gamma_e}{d_l}\sqrt{\frac{M}{2\pi R T_{sat}(p_l)}} \frac{\rho_g}{\rho_l - \rho_g}}_{\Re_e} L\alpha_l \rho_l \frac{T_{sat}(p_l) - T}{T_{sat}(p_l)} \end{cases} \tag{16}$$

In addition, volumetric phase change rate is:

$$J_{net} = J^+ + J^- \tag{17}$$

3.4 Entropy Generation Analysis Model

The irreversibility of DCC process consists of dissipations due to viscous and turbulence fluctuation, finite temperature difference in heat transfer as well as the phase change within the domain. From the view of irreversible thermodynamics, the transport equation for entropy in a Cartesian coordinate system can be written as tensor form [30]:

$$\frac{\partial}{\partial t}(\rho s)+\frac{\partial}{\partial x_i}(\rho_m s_m u_{i,m})=\frac{\partial J_i^s}{\partial x_i}+\dot{S}_{gen}^{'''} \quad (18)$$

where J_i^s is the entropy flux in i^{th} direction, and $\dot{S}'''_{gen}$ is the entropy generation rate.

The governing entropy equation for transport process including single-mixture flow, Fourier heat conduction, and incompressible fluid is considered to deduce the entropy generation model over the calculation domains in turbulent multiphase flows, as is in [31]:

$$\rho_m\left(\frac{\partial s_m}{\partial t}+u_{i,m}\frac{\partial s_m}{\partial x_i}\right)=\frac{\partial J_i^s}{\partial x_i}+\frac{\Phi}{T_m}+\frac{\Phi_{pc}}{T_m}+\frac{\Phi_\Theta}{T_m^2} \quad (19)$$

In the equation above, the last three terms on the right side are the entropy generation owing to viscous, phase change and the heat transfer with finite temperature difference, in proper order. Besides, the instantaneous variable can be divided into time-averaged part and fluctuating part in the RANS approach, such as:

$$s_m=\overline{s_m}+s'_m,\ T_m=\overline{T_m}+T'_m,\ u_{i,m}=\overline{u_{i,m}}+u'_{i,m},\ J_{net}=\overline{J_{net}}+J'_{net}$$

The terms in the Equation (19) are substituted by time-averaged variables, and the main processes are shown as follows.

3.4.1. Convective Terms

The time-averaged convective term for entropy on the left side of Equation (19) is written as:

$$\overline{\frac{\partial s_m}{\partial t} + u_{i,m}\frac{\partial s_m}{\partial x_i}} = \frac{\partial \overline{s_m}}{\partial t} + \overline{u_{i,m}}\frac{\partial \overline{s_m}}{\partial x_i} + \frac{\partial \overline{u'_{i,m} s'_m}}{\partial x_i} \tag{20}$$

3.4.2. Entropy Generation by Dissipation

Time-averaged entropy generation due to viscous dissipation contains two groups of terms, *i.e.*, averaged flow part and fluctuating part, which read as:

$$\frac{\overline{\Phi}}{T_m} = \frac{\mu_m}{\overline{T_m}}\left\{2\left(\frac{\partial \overline{u_{i,m}}}{\partial x_i}\right)^2 + \left(\frac{\partial \overline{u_{i,m}}}{\partial x_j} + \frac{\partial \overline{u_{j,m}}}{\partial x_i}\right)^2\right\} + \frac{\mu_m}{\overline{T_m}}\left\{2\left(\frac{\partial \overline{u'_{i,m}}}{\partial x_i}\right)^2 + \left(\frac{\partial \overline{u'_{i,m}}}{\partial x_j} + \frac{\partial \overline{u'_{j,m}}}{\partial x_i}\right)^2\right\} \tag{21}$$

in which, the T′m in the denominator is neglected [31].

3.4.3. Entropy Generation by Heat Transfer

In the time averaging process of entropy generation by heat transfer, the T′m is neglected for its existence in higher order terms. Then, the time-averaged entropy generation terms with respect to finite temperature gradients can be expressed as:

$$\frac{\overline{\Phi_\Theta}}{T_m^2} = \frac{\kappa_m}{\overline{T_m^2}}\left\{\left(\frac{\partial \overline{T_m}}{\partial x_i}\right)^2\right\} + \frac{\kappa_m}{\overline{T_m^2}}\left\{\left(\frac{\partial \overline{T_m{}'}}{\partial x_i}\right)^2\right\} \tag{22}$$

Here, the first term on the right side is the entropy generation due to heat transfer with time-averaged temperature gradients, while the second term on the right side is due to fluctuating temperature gradients.

3.4.4. Entropy Generation by Inner Phase Change

Considering the fluctuation of volumetric phase change rate, the time-averaged entropy generation by inner phase change is given by the following form:

$$\overline{\frac{\Phi_{pc}}{T_m}} = \frac{L \cdot \overline{J_{net}}}{\overline{T_m}} + \frac{L \cdot \overline{J'_{net}}}{\overline{T_m}} = \frac{L \cdot \overline{J_{net}}}{\overline{T_m}} \tag{23}$$

where L is the latent heat of water.

3.4.5. Time-Averaged Transport Equation for Entropy

Substituting the time-averaged terms into the Equation (19), the transport equation for the mean entropy in turbulence conditions leads to:

$$\begin{aligned}\left(\frac{\partial \overline{s_m}}{\partial t} + \overline{u_{i,m}} \frac{\partial \overline{s_m}}{\partial x_i}\right) = {} & \frac{\partial J_i^s}{\partial x_i} - \rho_m \frac{\partial \overline{u'_{i,m} s'_m}}{\partial x_i} + \underbrace{\frac{\mu_m}{\overline{T_m}} \left\{ 2\left(\frac{\partial \overline{u_{i,m}}}{\partial x_i}\right)^2 + \left(\frac{\partial \overline{u_{i,m}}}{\partial x_j} + \frac{\partial \overline{u_{j,m}}}{\partial x_i}\right)^2 \right\}}_{(\mathrm{I})} \\ & + \underbrace{\frac{\mu_m}{\overline{T_m}} \left\{ 2\overline{\left(\frac{\partial u'_{i,m}}{\partial x_i}\right)^2} + \overline{\left(\frac{\partial u'_{i,m}}{\partial x_j} + \frac{\partial u'_{j,m}}{\partial x_i}\right)^2} \right\}}_{(\mathrm{II})} + \underbrace{\frac{\kappa_m}{\overline{T_m^2}} \left(\frac{\partial \overline{T_m}}{\partial x_i}\right)^2}_{(\mathrm{III})} \\ & + \underbrace{\frac{\kappa_m}{\overline{T_m^2}} \overline{\left(\frac{\partial T_m'}{\partial x_i}\right)^2}}_{(\mathrm{IV})} + \underbrace{\frac{L \cdot \overline{J_{net}}}{\overline{T_m}}}_{(\mathrm{V})}\end{aligned} \tag{24}$$

Obviously, the term (II) and the term (IV) concerning the velocity and temperature fluctuations in different entropy generation sources above are still unclosed. Then, the information contained in the k-ε turbulence model can be used to replace the two terms in a closed form, which will be shown as follows:

(1) The exact turbulent dissipation approximately equals to the production of density ρ_m and the turbulent dissipation rate ε, therefore, the entropy generation rate due to turbulent dissipation reads as:

$$\dot{S}'''_{gen,D} = \frac{\overline{\rho_m \varepsilon}}{\overline{T_m}} \tag{25}$$

(2) Use the Boussinesque-like approach [31], and then the entropy generation due to fluctuating temperature gradients is:

$$\dot{S}'''_{gen,C'} = \frac{a_t}{a} \frac{\kappa_m}{\overline{T_m^2}} \left(\frac{\partial \overline{T_m}}{\partial x_i} \right)^2 \tag{26}$$

with a thermal diffusivity and a_t the turbulent thermal diffusivity. Assume that the turbulent thermal diffusivity is related as thermal diffusivity through:

$$\frac{\kappa}{a} = \frac{\kappa_t}{a_t} = \rho_m c_{pm} \tag{27}$$

Then, the entropy generation due to mean temperature gradients and entropy generation due to fluctuating temperature gradients can be combined as the entropy generation due to heat transfer:

$$\dot{S}'''_{gen,C} = \frac{(\kappa + \kappa_t)}{\overline{T_m^2}} \left(\frac{\partial \overline{T_m}}{\partial x_i} \right)^2 = \frac{\kappa_{eff}}{\overline{T_m^2}} \left(\frac{\partial \overline{T_m}}{\partial x_i} \right)^2 \tag{28}$$

From the derivation above, the local volumetric entropy generation is concluded as follows:

$$\dot{S}'''_{gen} = \underbrace{\frac{\kappa_{eff}}{\overline{T}_m^2}\left(\frac{\partial \overline{T_m}}{\partial x_i}\right)^2}_{heat\,transfer} + \underbrace{\frac{\mu_m}{\overline{T_m}}\left\{2\left(\frac{\partial \overline{u_{i,m}}}{\partial x_i}\right)^2 + \left(\frac{\partial \overline{u_{i,m}}}{\partial x_j} + \frac{\partial \overline{u_{j,m}}}{\partial x_i}\right)^2\right\}}_{viscous\ dissipation} +$$

$$\underbrace{\frac{\rho_m \varepsilon}{\overline{T_m}}}_{\substack{turbulent\\ dissipation}} + \underbrace{\frac{L \cdot \overline{J_{net}}}{\overline{T_m}}}_{inner\ phase\ change} \quad (29)$$

4. COMPUTATION SET-UP

4.1. Simulation Details

In the simulation, the SIMPLE algorithm [32] is used for pressure-velocity coupling. In addition, the second order upwind discretization scheme is used for solving the momentum equations and energy equation, while a first order upwind discretization scheme is used for solving turbulent kinetic energy equations, turbulent dissipation rate equations and volume fraction equations. Considering the instability resulting from two phase flow and condensation, the under-relaxation factors are initially set to be 0.1 for all variables at the first 0.1 s, and they subsequently change to be default value for the remainder of simulation time, while the time step is 0.001 s based on the previous work conducted by other researchers [5,6,7,9]. All the simulations were performed on a workstation with four Intel computer cores and 8 GB RAM.

As for the solutions residuals, two criteria are selected to determine whether they are convergent: (1) the residuals for all solutions except for energy solution should be below 1×10^{-3}, while the residual for solution of energy should be below 1×10^{-6}; (2) the entropy generation rate for all two phase domains hardly change with iterations in one time steps.

The boundary conditions are set as follows: the mass flux with value of 60 kg/m^{2}s for vapor is enforced at the inlet section, the temperature and pressure for vapor is 374.15 K and 3 kPa. The pressure outlet with value 0 Pa (gauge pressure) is employed for mixture at the outlet section. The rest boundaries are set to be adiabatic walls with no velocity slip.

The initial conditions are set as follows: inside the steam pipe is patched with vapor, the temperature and gauge pressure of which are 374.15 K and 3 kPa separately, whilst the temperature and gauge pressure of accumulated water inside the tank are 300 K and 0 Pa, and the operating pressure is 101.325 kPa, *i.e.*, 1 atm. For simplify, the phase change temperature is set to be constant, 373.15 K.

Steam in the fluid domain is treated as an incompressible ideal gas, which means that the density and specific heat capacity would change with temperature only, while other properties such as viscosity and thermal conductivity are constant value, which are taken from the FLUENT material database. By contrast, liquid water is regarded as incompressible fluid; and its properties are all specified as constant value, as is shown in Table 1. In the simulation, the values of evaporation/condensation coefficient are treated as constant, 0.01, derived from the [33,34].

4.2. Grid Independent Verification

In the current work, hexahedral elements are used for meshing, and grids with good quality are ensured using ICEM, a mesh generation tool. For the cylindrical pipe in the geometry, the block with O-Grid is adopted. The values of Y^+ is required to locate in the interval (11.225~200) for the adoption of standard wall functions, then the mesh near wall region is adapted to ensure $30 < Y^+ < 50$. Totally, three computational grids consisted with 38,6972, 633,452, and 914,597 nodes are constructed. Before the simulation, the effect of grid density on the results is investigated. Figure 3a,b show the transverse distribution of longitudinal velocity at selected locations (z_1 = 265 mm, z_2 = 280 mm) with the three different grids above. All three grids can predict similarly the velocity profile but the medium size mesh, *i.e.*, 633,452 nodes (≈ 80 × 80 × 100) is adopted considering the balance of accuracy and efficiency, as is shown in Figure 4.

5. RESULTS AND DISCUSSION

5.1. Verification & Validation

The method in the current study is based on the mixture model and the kinetic theory of gas, which is different from the previous work. To validate the current models, a simulation case with the same geometry as the model in the study by Takase *et al.* [35] has been performed. In this simulation, a velocity boundary which has a value of 25 m/s is adopted, the initial temperature of steam and accumulated water are 111 °C and 20 °C, respectively, and the gauge pressure inside the tank is 2.3 kPa. In addition, the steam changes to liquid water at 110.5 °C.

Figure 5 shows the comparisons of experimental results derived by Takase *et al.* [35] and the results by numerical predictions in the present work. Here, Figure 5a is the initial view, *i.e.*, at time t = 0.0 s, while Figure 5b is the result at time t= 0.5 s. From Figure 5b, it can be seen that the length, spreading angle and the width of plume jet are quite similar to those in the experimental observation, but the experiment shows a wavy and turbulent flow structure, while the shape of plume jet by present CFD prediction is smooth. The most possible reason is the adoption of RANS model, in which the turbulent kinetic energy formulation considers that all the normal components of stresses are isotropic, *i.e.*, the equal spreading of the plume jet in all directions. In general, the simulation steam plume shape agrees qualitatively well with the experimental observations.

Table 1. The thermophysical properties of fluids.

Phases	Density (kg/m^3)	Specific Heat Capacity (J/kg K)	Viscosity (Pa · s)	Thermal Conductivity (W/m K)
Vapor	Incompressible ideal-gas	Polynomial *	1.34×10^{-5}	0.0261
Water	998.2	4182	1.003×10^{-3}	0.6

* $c_p(T) = A_0 + A_1T + A_2T^2 + A_3T^3 + A_4T^4$, A_i ($i = 0,1,2,3,4$) are the default coefficients in ANSYS FLUENT.

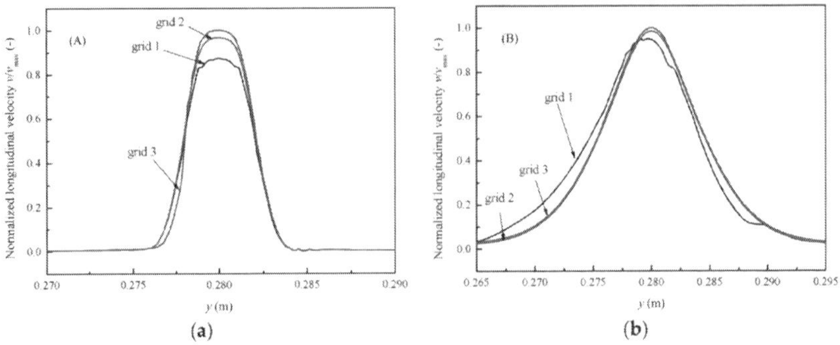

Figure 3. Transverse distribution for longitudinal velocity at selected locations (z_1 = 265 mm, z_2 = 280 mm) for different grid density (grid 1: black, 386,972 nodes; grid 2: red, 633,452 nodes; grid 3: blue, 914,597 nodes. v_{max} is the max velocity among the three grids).

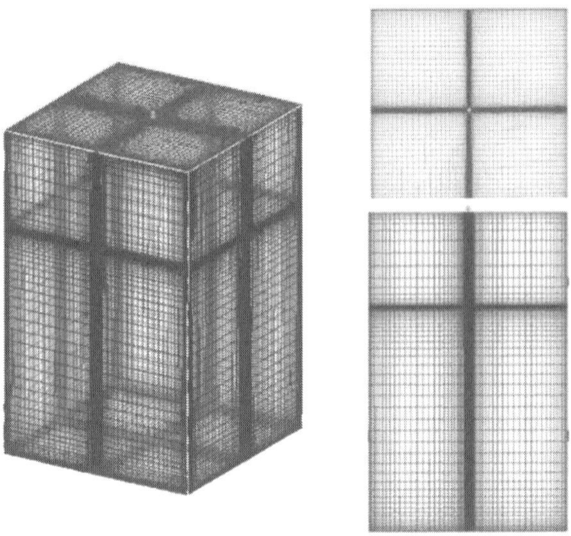

Figure 4. Schematic view of simulation mesh of the hexahedral tank.

Figure 6 shows the comparisons of transverse temperature distribution at selected longitudinal location between the present numerical predictions and the published work by Takase *et al.* [35], where X^+ is the dimensionless transverse location, and Z^+ is the dimensionless distance departure from the steam pipe exit ($X^+ = x/d_0$, $Z^+ = z/d_0$, d_0 is the inner diameter of steam pipe). Both the results show steep value in the vicinity of the jet, and smooth temperature change in the outward zone. Besides, the CFD prediction agrees quantitatively well with the numerical results derived by Takase *et al.* [35] (within 5%) in the central location, whilst it follows the same trend of the results by Takase *et al.* [35] with marginal deviation within 14% in the outer location.

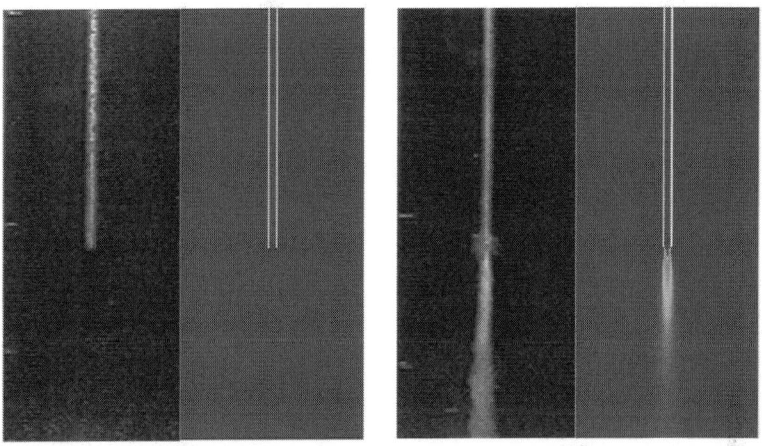

(a) (b)

Figure 5. The comparisons of simulation steam shape and the experimental observations at different time; (**a**) t = 0.0 s; (**b**) t = 0.5 s.

5.2. Numerical Results

5.2.1. Velocity Profile

Figure 7 shows the velocity streamline in the x = 0 plane at different time, t = 4 ms, t = 8 ms, t = 12 ms, t = 16 ms, t = 44 ms, and t = 120 ms. From the figures, it can be seen that, once the steam discharging into the accumulated water, the interaction between steam and water due to momentum exchange decays the velocity of steam to 0 immediately in certain distance, which drives the surrounding water flow slightly. At the jet interface, surrounding water is entrained into the plume region due to buoyancy effect, and then a recirculation zone as shown in figures is formed. In this way, intense circulation in the tank enhance the thermal mixing effect during DCC processes.

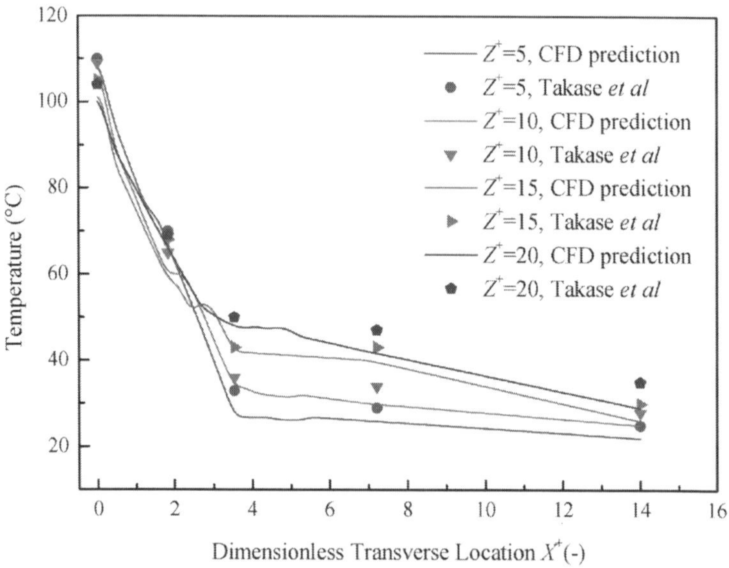

Figure 6. The comparisons of the transverse temperature distribution at selected longitudinal locations between published data and present CFD predictions.

Transverse profile of V_z for selected longitudinal positions (z = 0.26 m and z = 0.30 m) at different time are displayed in Figure 8, which indicate that V_z of steam in the central location near pipe exit decreases rapidly after it reaches its maximum value. The observations correspond to the results obtained by Dahikar *et al.* [7] in their experiments and CFD simulation. Besides, it can be observed that the velocity in the centerline is not at the maximum all the time, and sometimes it increases in the jet zone deviated from the center line. Furthermore, the velocity changes sharply with the variation of transverse location near the region of pipe exit, while it changes smoothly in the outward zone, due to finite spreading width of jets in the pool water.

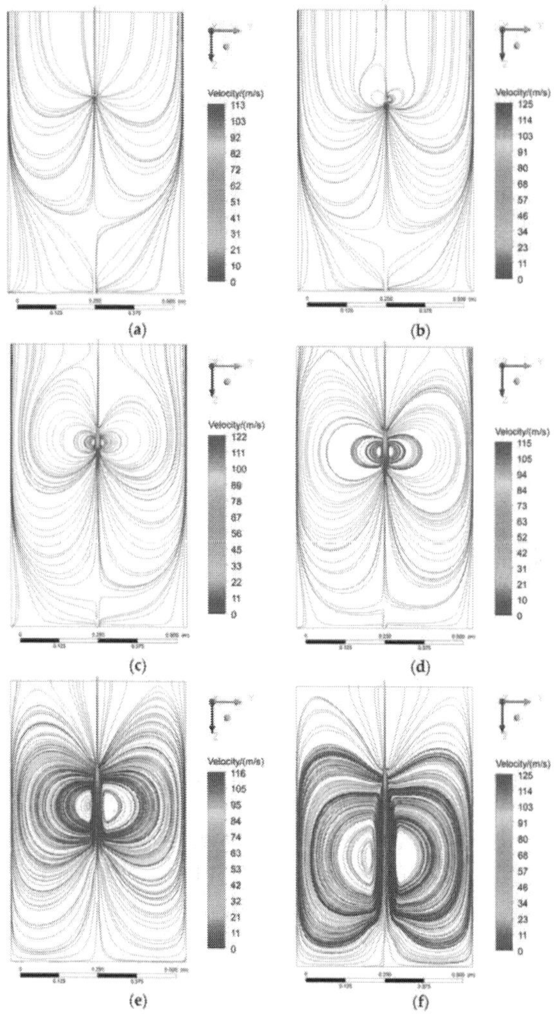

Figure 7. Velocity streamline in the $x = 0$ plane at different time. (**a**) $t = 4$ ms; (**b**) $t = 8$ ms; (**c**) $t = 12$ ms; (**d**) $t = 16$ ms; (**e**) $t = 44$ ms; (**f**) $t = 120$ ms.

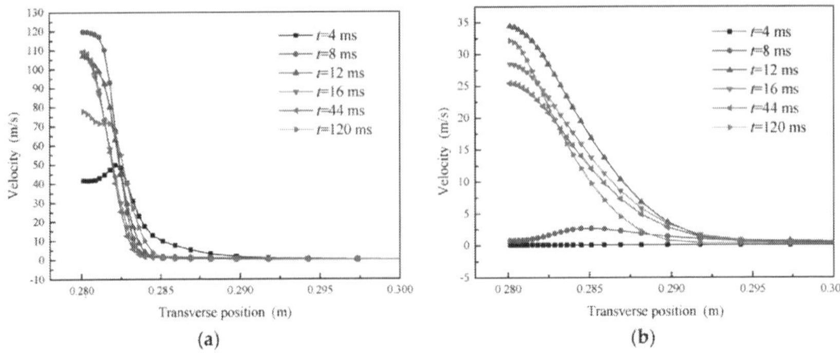

Figure 8. Transverse profile of V_z for selected longitudinal position at different time; (**a**)z = 0.26 m; (**b**) z = 0.30 m.

5.2.2. Temperature Field

The heat transfer process is modeled by associating the energy source with phase change in control volume, and transient simulation is carried out to determine the temperature field. Predictions of temperature in the x = 0 plane at different time and transverse profile of temperature for selected longitudinal position at different time are shown in Figure 9 and Figure 10, respectively. From the Figure 9, it can be observed that steam temperature in the pipe is almost a constant, 374.15 K, and decreases dramatically when the steam is discharged into the water, due to the energy exchange between steam and pool water. Large temperature variation during the DCC process is characterized by Figure 9a–d, while the predictions in Figure 9e–f indicate that temperature field varies little in the following time. The minor temperature change of pool water is due to low mass flux (60 kg/m^2 s) and low temperature (374.15 K) of steam.

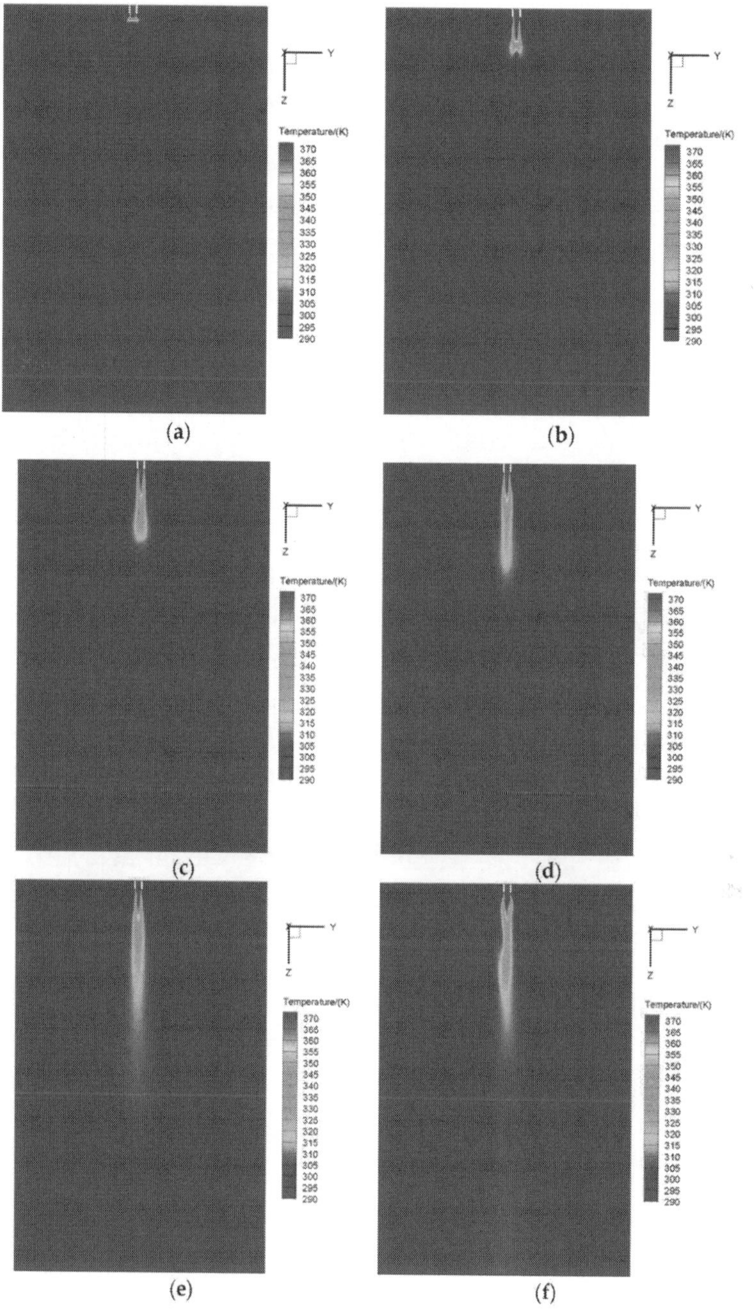

Figure 9. Contours plot of temperature in the $x = 0$ plane at different times. (**a**) $t = 4$ ms; (**b**) $t = 8$ ms; (**c**) $t = 12$ ms; (**d**) $t = 16$ ms; (**e**) $t = 44$ ms; (**f**) $t = 120$ ms.

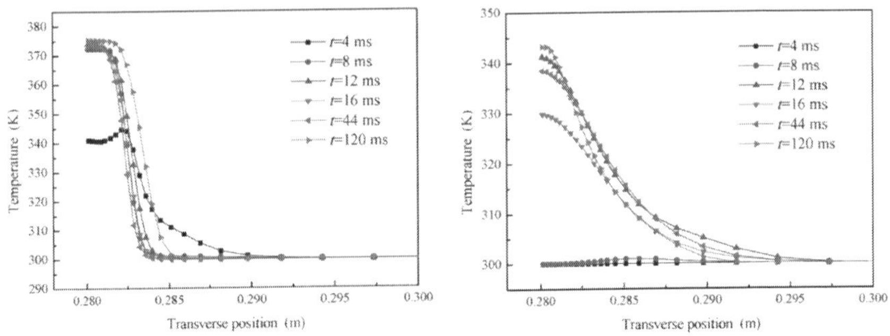

(a) (b)

Figure 10. Transverse profile of temperature for selected longitudinal position at different time; (**a**) $z = 0.26$ m; (**b**) $z = 0.30$ m.

In Figure 10, a similar transverse profile of temperature can be observed, which indicates that the temperature field (heat transfer) is relevant to the velocity profile. In the central location of steam jets, the entrainment of subcooled water into the plume generates small perturbation, resulting in lower temperature in the plume, e.g., $z = 0.26$ m at $t = 4$ ms, $z = 0.30$ m at $t = 8$ ms. Moreover, a sharp decrease of temperature occurs at the vapor-liquid interface, while smooth decrease of temperature within 5 K is observed far away from pipe exit, e.g., liquid region. The results of numerical predictions show the similar profile as the work by Gulawani [5,6].

5.2.3. Plume Shape

A vapor cavity called plume jet can be observed in the region near pipe exit after the steam ejecting from the nozzle.Figure 11 shows the contours plot of vapor void fraction in the $x = 0$ plane at different time. According to the plume shapes shown in the figure, three stages can be discriminated, *i.e.*, the initial stage (Figure 11a,b), developing stage (Figure 11c–e) and the oscillatory stage (Figure 11f). In the initial stage, the steam spreads immediately after discharged into the subcooled water, and the plume shows no fixed shape; then in the developing stage, the plume begin to change its form from "pear" shape to elliptical boundary, and the size of plume grows gradually; in the oscillatory stage, the small perturbation makes the plume shape become ellipsoidal shape with wavy boundary.

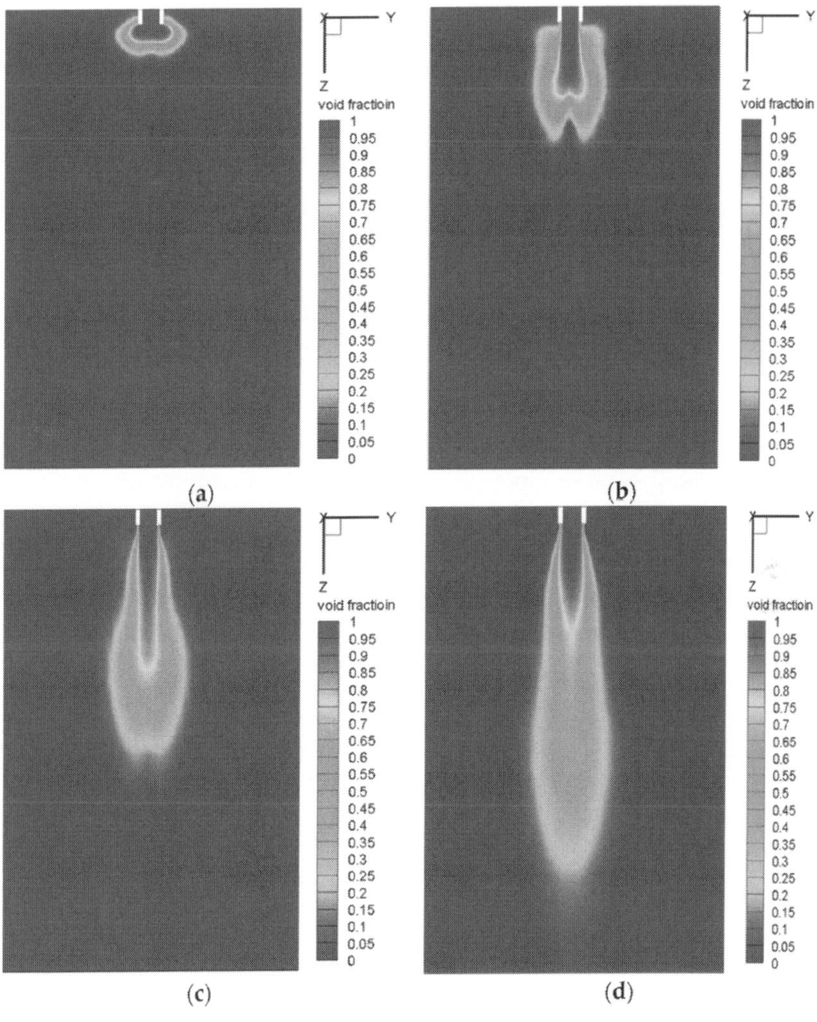

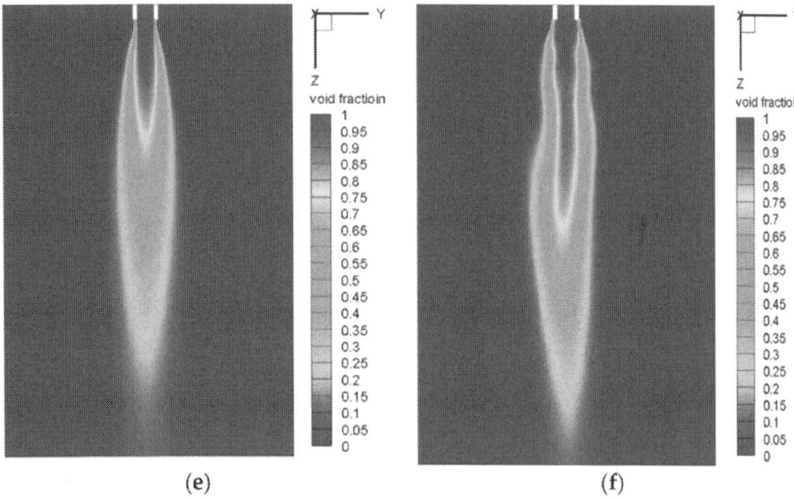

Figure 11. Contours plot of vapor void fraction in the $x = 0$ plane at different time. (**a**) $t= 4$ ms; (**b**) $t = 8$ ms; (**c**) $t = 12$ ms; (**d**) $t = 16$ ms; (**e**) $t = 44$ ms; (**f**) $t = 120$ ms.

There are some possible reasons to explain the phenomenon: (1) steam is discharged into the stagnant water at high speed, and strong fluctuation occurs in the initial stage, which prevents the plume from forming a fixed shape; (2) with the proceeding of steam condensation, the surrounding water is heated up and the degree of water subcooling decreases, which results in the growth of plume; (3) simultaneous decrease of fluctuation and subcooling make the plume grow faster, while instantaneous condensation gives rise to the formation of oscillation.

5.2.4. Mass Transfer

Figure 12 presents the instantaneous condensation rate in the DCC process of steam. If the boundary with vapor void fraction 0.5 is defined as vapor-liquid interface, it is found that steam mainly condenses at the interface compared with the contour plot of volume fraction of vapor, as is shown in Figure 11. The condensation rate in the numerical simulation is rather small, and the possible reason may be that the low steam mass flux generates less turbulent fluctuation, which obviously weakens the effect of thermal mixing associating with the fluctuation.

In addition, the condensation rate in the initial stage is larger than the condensation rate in other stages, due to gradual decreasing subcooling of pool water.

5.3. Entropy Generation

The volumetric entropy generation rate (entropy generation rate per unit volume) is determined in this section, based on basic parameters required. Figure 13 shows the contour plot of total entropy generation rate (EGR) per unit volume in the x= 0 plane at different time. As is shown in Figure 13, two strips with larger value of the volumetric EGR occur in the vicinity of steam pipe exit. Upon the steam discharging into the stagnant water, there is no certain boundary to restrict the steam flow. From the velocity streamline plot (Figure 7), it can be found that the large velocity gradients exist near the pipe exit. And moreover, to the best of our knowledge, the turbulence fluctuations will appear at the vapor-liquid interface due to the Kelvin-Helmholtz instabilities. These two terms lead to great dissipation initially. Compared with the other stages, the local value of volumetric EGR in the initial stage is much larger, while the region possessing the considerable EGR is smaller due to the shorter penetration length and finite spreading width of the plume jet, as is shown in Figure 11. In the developing stage, the values of volumetric EGR reduces shaply, and the zones with lagre volumetric EGR continues to grow, which corresponds to the behaviours of plume jet in this stage. In the oscillatory stage, the profile of local EGR becomes asymmetrical owing to the perturbation of flow and heat transfer stated in the previous sections.

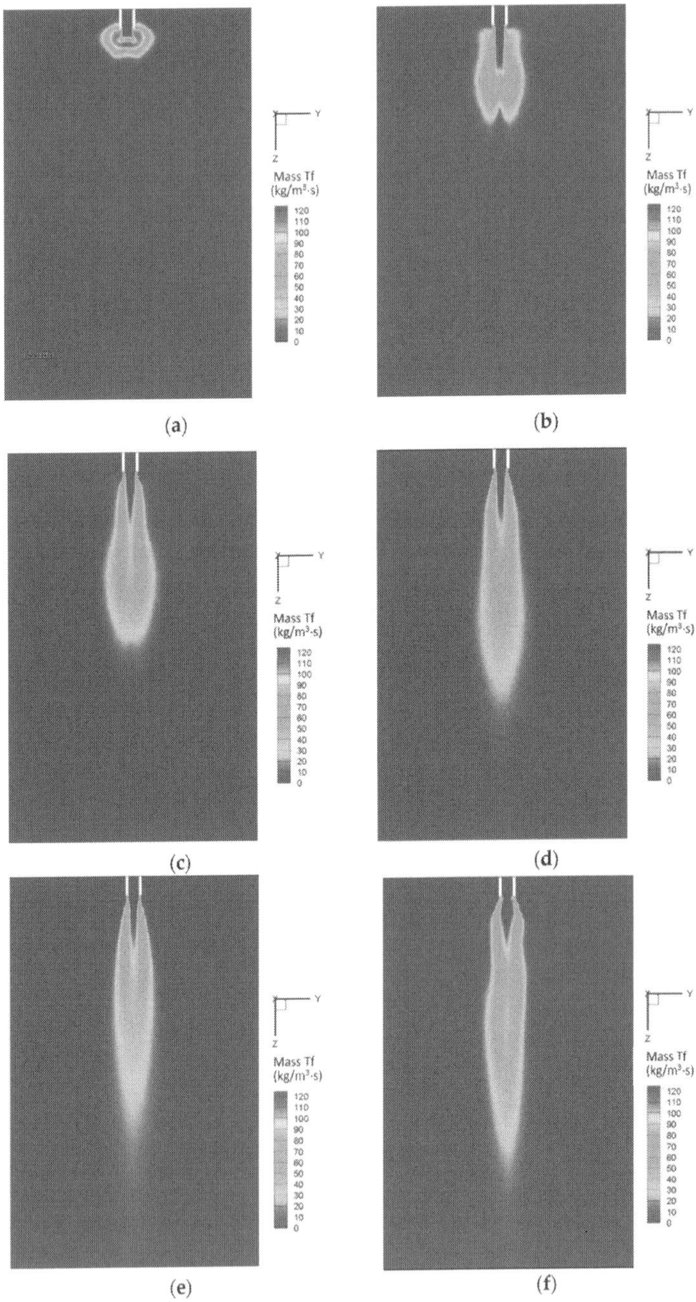

Figure 12. Instantaneous condensation rate in the x = 0 plane at different time. (**a**) t = 4 ms; (**b**) t = 8 ms; (**c**) t = 12 ms; (**d**) t = 16 ms; (**e**) t = 44 ms; (**f**) t = 120 ms.

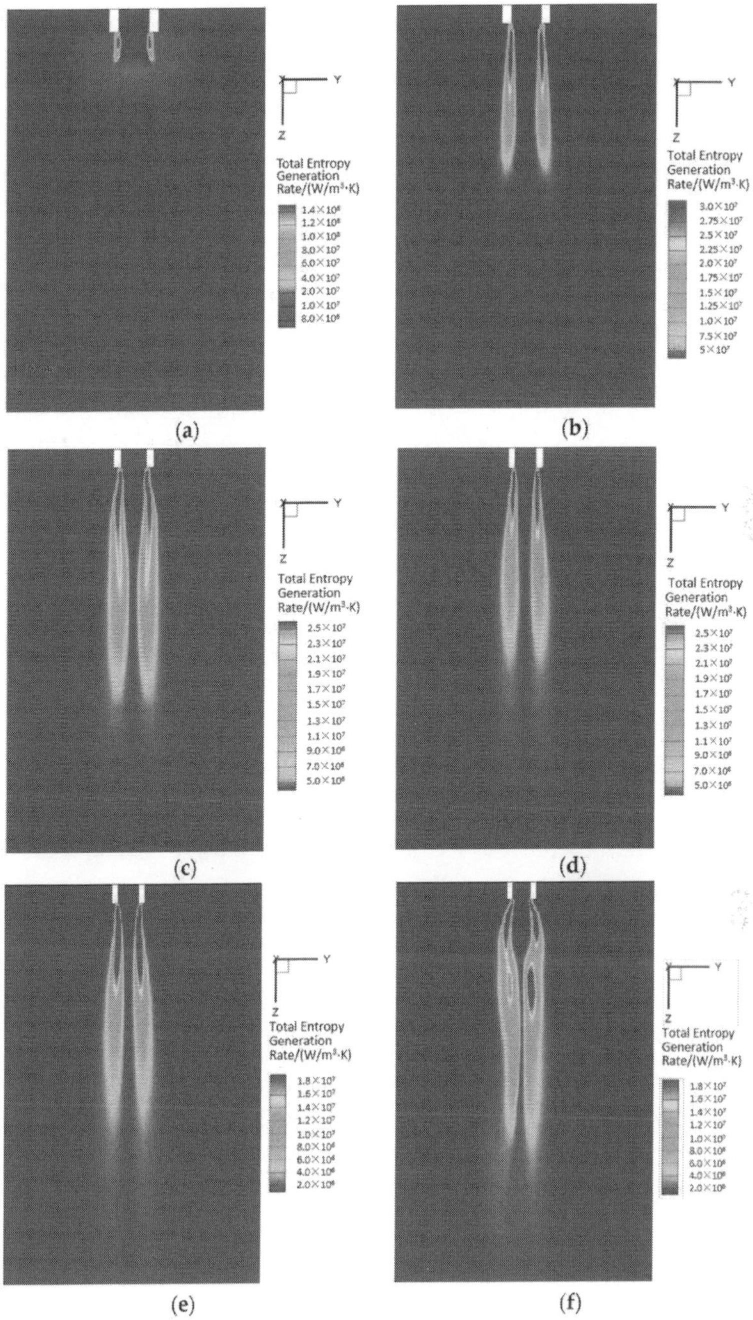

Figure 13. Total EGR per unit volume in the $x = 0$ plane at different time. (**a**) $t = 4$ ms; (**b**) $t = 8$ ms; (**c**) $t = 12$ ms; (**d**) $t = 16$ ms; (**e**) $t = 44$ ms; (**f**) $t = 120$ ms.

The volumetric EGR is integrated to the whole two phase flow domain V, which gives the entropy generation rate: representing the total dissipation in the whole domain.

$$\dot{S}_{gen} = \int_V \dot{S}_{gen}^{'''} \mathrm{d}V \tag{30}$$

Table 2 has listed out the EGR due to four kinds of irreversibility and the total EGR, which are used for describing the energy dissipation at different time during the DCC processes. From the table, it can be seen that the total EGR is higher initially for the violent turbulence, and declines quickly in the following time, accompanying some oscillation. In the process, the absolute value of EGR resulting from turbulence dissipation decreases apparently, while the EGR on account of heat transfer and inner phase change increase. The possible reason accounting for the phenomenon may be that the heat transfer and phase change can not response in time when the steam is discharged into the stagnant water. The DCC process performs turbulence dominant firstly. And subsequently, the heat transfer and phase change start gradually. Moreover, the value volumetric EGR due to viscous dissipation is relatively low during the whole process. The decrease of EGR proves that the process is becoming more and more energy-economical.

Figure 14 shows the variation of contributions to total EGR with time resulting from the four kinds of irreversibility. In the figure, the EGR due to turbulence fluctuation occupies the largest proportion among the four irreversibility element in the initial stage, then decreases sharply. The effects of heat transfer irreversibility and inner phase change irreversibility on the EGR increase, which makes DCC process become heat dominant in the developing stage. In the oscillatory stage, the contributions of four kinds of irreversibility to total EGR hardly changes with time in the macroscopic view, and the process is still heat dominant. However, some oscillation still exists here if a more detailed look was taken. From begining to the end, the effect of viscous dissipation is minor. The reason is assumed that: the process is a phenomenon with violent perturbation, and mean flow element has lost its significance among all irreversibility, which is a common view for steam jet in many researches. The results here are accordance with the derivation of local volumetric EGR above. The analysis above can contribute to understand and quantify the major reversibility during the whole process.

From the table and figures, the three stages can be clearly discriminated according to different EGR characteristics and their partial proportions with time, which may be difficultly distinguished from the thermal-hydraulic information, such as velocity streamline and temperature profile solely. Based on the EGR approach, the dissipation characteristics of the DCC is derived, which may contribute to better understand the process. Then, it may also contribute to optimize the design for facilities, including pipe diameter and pipe length, to minify the size of two strips of volumetric EGR. In addition, it may help optimization of configuration for operation conditions, including temperature of water and mass flux of steam, to reduce the value of EGR. Moreover, other factors such as the pipe with or without diffusor can be investigated using the EGR assessment approach for thermal mixing analysis concerning DCC phenomenon. The discussions will be further conducted in future works.

6. CONCLUSIONS

In the current work, the DCC process of steam jet in a subcooled water tank is investigated through numerical simulation and the entropy generation assessment model based on the local equilibrium hypothesis is developed to evaluate the energy dissipation characteristics. The main inferences can be summarized as follows:

(1) A different mathematical model from the previous studies by other researchers is proved to be valid through the comparisons with the results obtained by Takase *et al.* [35], Dahikar *et al.* [7] and Gulawani [5,6].

Table 2. Entropy Generation Rate (EGR) in DCC process.

Time/(ms)	EGR_Heat Transfer/(W/K)	EGR_Viscous/ (W/K)	EGR_Turbulence/ (W/K)	EGR_Inner Phase Change/(W/K)	Total EGR/ (W/K)
2	0.04728	0.00016	12750.01000	0.74853	12750.8080
4	0.52859	0.00031	1361.27000	2.90208	1364.7051
8	12.97281	0.00090	109.47290	5.96174	128.4084
12	27.41757	0.00099	28.85747	8.35380	64.6298
16	29.43516	0.00087	12.32487	9.37835	51.1393
20	31.36492	0.00103	7.30897	6.73753	45.4124
24	30.30458	0.00086	4.67426	9.81291	44.7926
28	30.89448	0.00086	3.51669	9.73798	44.1500
32	31.55728	0.00087	2.87975	9.67914	44.1170
36	32.04326	0.00087	2.49488	9.65655	44.1956
40	32.34219	0.00087	2.24267	9.65505	44.2471
44	32.49960	0.00087	2.06793	9.66349	44.2319
64	31.01836	0.00079	0.89096	9.67850	41.5886
140	32.90304	0.00065	0.53704	9.93461	43.3753
180	30.86176	0.00062	0.42636	9.48106	40.7698

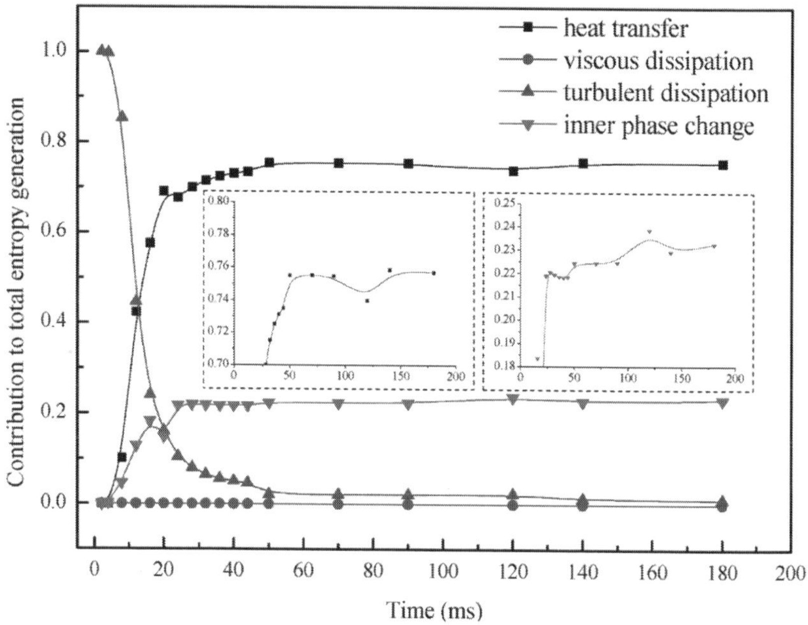

Figure 14. The variation of contributions of four kinds of irreversibility to total entropy generation with time.

(2) Three distinct stages of DCC are discriminated clearly at the present conditions, *i.e.*, initial stage, developing stage and oscillatory stage. In the initial stage, the plume shows no fixed shape. In the developing stage, the plume begins to act as an elliptical boundary, and the size of the plume grows quickly. In the oscillatory stage, the plume shape becomes ellipsoidal shape with disturbed structure.

(3) The local volumetric EGR in the initial stage is much larger than those in other stages, but the region possessing considerable entropy generation rate is smaller than other stage. The decrease of EGR proves that the process conform to increasingly economical energy utilization.

(4) The largest proportion in total EGR is occupied by turbulence fluctuation in the initial stage, and then it decreases apparently in

the following time, meanwhile, the contributions of heat transfer irreversibility and inner phase change irreversibility to the local entropy generation increase, which makes DCC process become heat dominant in the developing and the oscillatory stage. The variation of EGR can be used to characterize the the dissipation and proceeding of DCC process.

In general, the current work presents a promising method to investigate the DCC process, in which an analytical solution to determine condensation and evaporation rate is introduced. And furthermore, it also proposes an effective approach based on the entropy generation for better understanding the DCC process, including the the thermal hydraulic and the energy dissipation information, which can contribute to design the thermodynamically efficient equipment involved in the DCC process in future works.

ACKNOWLEDGMENTS

This work is supported by the Foundation for Innovative Research Groups of the National Natural Science Foundation of China (Grant No. 51421063) and the National Natural Science Foundation of China (Grant No. 51536001).

AUTHOR CONTRIBUTIONS

The research concept and research design was completed by Hao-Chun Zhang and Yu Ji initially. Yu Ji conducted the calculation under the help of Jian-Fei Tong. Yu Ji and Yi-Ning Zhang undertook the data analysis. The manuscript was primarily written by Yu Ji, with the assistance of Hao-Chun Zhang, Xu-Wei Wang and Han Wang. All authors have read and approved the final manuscript.

REFERENCES

1. Kerney, P.J.; Faeth, G.M.; Olson, D.R. Penetration characteristics of a submerged steam jet. *AIChE J.* **1972**, *18*, 548–553.

2. Weimer, J.C.; Faeth, G.M.; Olson, D.R. Penetration of vapor jets submerged in subcooled liquids. *AIChE J.* **1973**, *19*, 552–558.

3. Aya, I.; Nariai, H. Evaluation of heat-transfer coefficient at direct-contact condensation of cold water and steam. *Nucl. Eng. Des.* **1991**, *131*, 17–24.

4. Chan, C.K.; Lee, C.K.B. A regime map for direct contact condensation. *Int. J. Multiph. Flow* **1982**, *8*, 11–20.

5. Gulawani, S.S.; Joshi, J.; Shah, M.; RamaPrasad, C.S.; Shukla, D.S. CFD analysis of flow pattern and heat transfer in direct contact steam condensation. *Chem. Eng. Sci.* **2006**, *61*, 5204–5220.

6. Gulawani, S.S.; Dahikar, S.K.; Mathpati, C.S.; Joshi, J.B.; Shah, M.S.; RamaPrasad, C.S.; Shukla, D.S. Analysis of flow pattern and heat transfer in direct contact condensation. *Chem. Eng. Sci.* **2009**, *64*, 1719–1738.

7. Dahikar, S.K.; Sathe, M.J.; Joshi, J.B. Investigation of flow and temperature patterns in direct contact condensation using PIV, PLIF and CFD. *Chem. Eng. Sci.* **2010**, *65*, 4606–4620.

8. Shah, A.; Chughtai, I.R.; Inayat, M.H. Numerical simulation of direct-contact condensation from a supersonic steam jet in subcooled water. *Chin. J. Chem. Eng.* **2010**, *18*, 577–587.

9. Li, S.Q.; Wang, P.; Lu, T. Numerical simulation of direct contact condensation of subsonic steam injected in a water pool using VOF method and LES turbulence model. *Prog. Nucl. Energy* **2015**, *78*, 201–215.

10. Torabi, M.; Zhang, K. Classical entropy generation analysis in cooled homogenous and functionally graded material slabs with variation of internal heat generation with temperature, and convective-radiative boundary conditions.*Energy* **2014**, *65*, 387–397.

11. Hassan, M.; Sadri, R.; Ahmadi, G.; Ahmadi, G.; Dahari, M.B.; Kazi, S.N.; Safaei, M.R.; Sadeghinezhad, E. Numerical study of entropy generation in a flowing nanofluid used in micro-and minichannels. *Entropy* **2013**, *15*, 144–155.

12. Makinde, O.D.; Eegunjobi, A.S. Entropy generation in a couple stress fluid flow through a vertical channel filled with saturated porous media. *Entropy* **2013**, *15*, 4589–4606.

13. Seo, J.W.; Kim, Y.H.; Kim, D.; Choi, Y.D.; Lee, K.J. Heat Transfer and Pressure Drop Characteristics in Straight Microchannel of Printed Circuit Heat Exchangers. *Entropy* **2015**, *17*, 3438–3457.

14. Cho, C.C.; Yau, H.T.; Chiu, C.H.; Chiu, K.C. Numerical Investigation into Natural Convection and Entropy Generation in a Nanofluid-Filled U-Shaped Cavity. *Entropy* **2015**, *17*, 5980–5994.

15. Bejan, A. *Entropy Generation Minimization: The Method of Thermodynamic Optimization oF Finite Size System and Finite-Time Processes*; CRC Press: Boca Raton, FL, USA, 1995.

16. Bejan, A. *Entropy Generation Through Heat and Fluid Flow*; Wiley: New York, NY, USA, 1982.

17. Vargas, J.V.C.; Bejan, A. Thermodynamic optimization of the match between two streams with phase change. *Energy***2000**, *25*, 15–33.

18. Goudarzi, N.; Talebi, S. An approach to stability analysis and entropy generation minimization in the single-phase natural circulation loops. *Energy* **2015**, *80*, 213–226.

19. Lior, N.; Sarmiento, D.W.; Al-Sharqawi, H.S. The exergy fields in transport processes: Their calculation and use.*Energy* **2006**, *31*, 553–578.

20. Adeyinka, O.B.; Naterer, G.F. Optimization correlation for entropy production and energy availability in film condensation. *Int. Commun. Heat Mass Transf.* **2004**, *31*, 513–524.

21. Thiel, G.P.; Lienhard, J.H. Entropy generation in condensation in the presence of high concentrations of noncondensable gases. *Int. J. Heat Mass Transf.* **2012**, *55*, 5133–5147.

22. Revellin, R.; Lips, S.; Khandekar, S.; Bonjour, J. Local entropy generation for saturated two-phase flow. *Energy* **2009**,*34*, 1113–1121.

23. Herwig, H.; Schmandt, B. How to Determine Losses in a Flow Field: A Paradigm Shift towards the Second Law Analysis. *Entropy* **2014**, *16*, 2959–2989.

24. Wenterodt, T.; Herwig, H. The Entropic Potential Concept: A New Way to Look at Energy Transfer Operations.*Entropy* **2014**, *16*, 2071–2084.

25. Fluent Inc. *Fluent 14.0 Theory Guide*; Fluent Inc.: Pittsburgh, PA, USA, 2012.

26. Marek, R.; Straub, J. Analysis of the evaporation coefficient and the condensation coefficient of water. *Int. J. Heat Mass Transf.* **2001**, *44*, 39–53.

27. Zhang, X.B.; Li, J.F.; Zhu, J.K.; Qiu, L.M. Computational fluid dynamics study on liquefied natural gas dispersion with phase change of water. *Int. J. Heat Mass Transf.* **2015**, *91*, 347–354.

28. Knudsen, M. Die maximale verdampfungsgeschwindigkeit des quecksilbers. *Ann. Phys.* **1915**, *352*, 697–708.

29. Pru, W. Verdampfungsvorgang. *Forsch. Gebiete Ing.* **1941**, *12*, 258–260.

30. Jou, D.; Casas-Vázquez, J.; Lebon, G. *Extended Irreversible Thermodynamics*; Springer: Berlin/Heidelberg, Germany, 1996.

31. Kock, F.; Herwig, H. Local entropy production in turbulent shear flows: A high-Reynolds number model with wall functions. *Int. J. Heat Mass Transf.* **2004**, *47*, 2205–2215.

32. Patankar, S.V. *Numerical Heat Transfer and Fluid Flow*; McGraw-Hill: New York, NY, USA, 1981.

33. Narusawa, U.; Springer, G.S. Measurement of evaporation rates of water. *J. Colloid Interface Sci.* **1975**, *50*, 392–395.

34. Nabavian, K.; Bromley, L.A. Condensation coefficient of water. *Chem. Eng. Sci.* **1963**, *18*, 651–660.

35. Takase, K.; Ose, Y.; Kunugi, T. Numerical study on direct-contact condensation of vapor in cold water. *Fusion Eng. Des.* **2002**, *63*, 424–428.

10

TWO-PHASE NUMERICAL MODEL FOR THERMAL CONDUCTIVITY AND CONVECTIVE HEAT TRANSFER IN NANOFLUIDS

Sasidhar Kondaraju and Joon Sang Lee

Department of Mechanical Engineering, Yonsei University, Seoul, Korea

ABSTRACT

Due to the numerous applications of nanofluids, investigating and understanding of thermophysical properties of nanofluids has currently become one of the core issues. Although numerous theoretical and numerical models have been developed by previous researchers to understand the mechanism of enhanced heat transfer in nanofluids; to the best of our knowledge these models were limited to the study of either thermal conductivity or convective heat transfer of nanofluids. We have developed a numerical model which can estimate the enhancement in both the thermal conductivity and convective heat transfer in nanofluids. It also aids in understanding the mechanism of heat transfer enhancement. The study reveals that the nanoparticle dispersion in fluid medium and nanoparticle heat transport phenomenon are equally important in enhancement of thermal conductivity. However, the enhancement in convective heat transfer was caused mainly due to the nanoparticle heat transport mechanism. Ability of this model to be able to understand the mechanism of convective heat transfer enhancement distinguishes the model from rest of the available numerical models.

1. BACKGROUND

The thermal conductivity of thermofluid plays an important role in the development of energy-efficient heat transfer equipment. Passive enhancement methods are commonly utilized in the electronics and transportation devices, but the thermal conductivity of the working fluids such as ethylene glycol (EG), water and engine oil is relatively lower than those of solid particles. In that regard, the development of advanced heat transfer fluids with higher thermal conductivity is in a strong demand.

To obtain higher thermal conductivity, numerous theoretical and experimental studies of the effective thermal conductivity of solid-particle suspensions have been conducted dated back to the classic work of Maxwell [1]. The key idea was to exploit the very high thermal conductivity of solid particles, which can be hundreds and even thousands of times greater than that of the conventional heat transfer fluids such as ethylene glycol and water, but most of these studies were confined to suspensions of millimeter- and micrometer-sized particles [2, 3]. Although such suspensions show higher thermal conductivity, they suffer from stability problems. In particular, particles tend to settle down very quickly and thereby causing severe clogging [4].

Unlike macro- and microparticles suspended in fluid, applications of nanoparticles provide an effective way of improving heat transfer characteristics of fluids. Particles, which are smaller than 100 nm in diameter exhibit properties different from those of microsized particles. It was demonstrated that nanofluids are extremely stable and exhibit no significant settling under static conditions [4, 5]. From previous investigations [6, 7, 8, 9, 10, 11], it was also observed that nanofluids exhibit substantially higher thermal conductivity even at very low volume concentrations ($\Phi < 0.05$) of suspended nanoparticles.

Ever since it was observed that nanofluids showed an improved thermal conductivity, researchers have tried to develop numerical models to predict and understand the heat transfer mechanism in nanofluids accurately. Bhattacharya et al. [12] and Jain et al. [13] performed Brownian dynamic simulations to predict the thermal conductivity enhancement in nanofluids. Xuan and Yao [14] developed a lattice Boltzmann model to investigate the nanoparticle distribution in stationary fluid. Evans [15] and Sarkar and Selvam [16] have used molecular dynamics simulations to predict the thermal conductivity in nanofluids. Molecular dynamics simulations were performed at very small volume fractions or in highly idealized conditions and thus could not be validated with the experimental data. Simulation of naturalistic data would have necessitated a large computational power which is beyond the scope of current computers. To avoid this, the Brownian dynamics simulations omit fluid molecules and add the effect of hydrodynamic interactions by including position-dependent interparticle friction tensor. The above models can only be used to simulate the still fluid conditions and cannot be used to predict the convective heat transfer enhancement in nanofluids. To predict the convective heat transfer in nanofluids, Maiga et al. [17] performed numerical simulations using a single-phase Navier-Stokes model. The physical properties of nanofluids (density, thermal conductivity and viscosity) were predicted by assuming that the nanoparticles were well dispersed in the base fluid. The model cannot explain the mechanism of convective heat transfer enhancement in nanofluids because of the fact that the model is based on single-phase flow assumption. In the present study, a two-phase model is being considered. In this model, fluid properties are modified due to the dispersion of particles in the fluid medium and due to the interfacial interaction between particles and fluid. Thus, the need of correlation equations for predicting the change in fluid properties due to the presence of nanofluids can be evaded.

2. MATHEMATICAL MODEL

In the present study, an Eulerian-Lagrangian two-phase flow model is discussed, and the model is used to predict thermal conductivity and convective heat transfer enhancements in nanofluids. The model also gives an insight into the mechanism of heat transfer enhancements. The numerical model used in the present study solves for multiphase Navier-Stokes equations, where fluid phase is solved in Eulerian reference frame and particle phase is solved in Lagrangian reference frame. A brief overview of the model is presented in this paper. Readers are referred to S Kondaraju et al. [18] detailed information on the model.

In the Lagrangian frame of reference, the equation of motion of nanoparticle and time-dependent particle temperature equation are given by,

$$(\mathrm{d}x_i^n)/\mathrm{d}t \equiv \mathrm{v}_i^n \tag{1}$$

$$\frac{\mathrm{d}v_i}{\mathrm{d}t} = F_{\mathrm{Di}} + F_{\mathrm{Bi}} + F_{\mathrm{Ti}} + F_{\mathrm{Vi}} \tag{2}$$

$$\frac{\mathrm{d}T_{\mathrm{p}}}{\mathrm{d}t} = \frac{\mathrm{Nu}}{\tau_{\mathrm{T}}} \frac{(\theta_{\mathrm{f}} - T_{\mathrm{p}})}{2} \tag{3}$$

Dispersion of nanoparticles was modeled by applying hydrodynamic drag force (F_{Di}) [19], Brownian force (F_{Bi}) [20], thermophoresis force (F_{Ti}) [21] and van der Waals force (F_{Vi}) [22] in the nanoparticle momentum equation. The coagulation of nanoparticles was also controlled by the van der Waals force acting on the adjacent nanoparticles. A cutoff distance of 0.2 nm was implemented in calculation of the van der Waals force. When the distance between the particles is less than the cutoff distance, particles were modeled to coagulate into one sphere with diameter equal to the summation of diameters of two coagulated particles. x_i^n and v_i^n are the instantaneous particle position and velocity of the n th particle, respectively. Subscript i represents the tensor notation. τ_T is thermal response time of the particle and given as $\tau_T = \frac{\rho_p c_p d_p^2}{12 k_f}$. k_f, d_p, c_p and ρ_p are the thermal conductivity of the base fluid, diameter, specific heat and density of the particle, respectively. **Nu** is the Nusselt number. θ_f is the fluid fluctuation temperature in the neighborhood of the particle and T_p is the temperature of the particle. It should be noted that in the present coagulation model the volume of coagulated particles is greater than the volume of particles when they coagulate in a real world situation (due to the assumption that two coagulated particles have a diameter equal to the summation of diameters of the two particles). However, the maximum increase in the volume concentration over time has been calculated and has been found to be of negligible amount to make any significant difference to the present results (see Appendix for the calculation).

Time-dependent, three-dimensional Navier-Stokes equations are solved in a cubical domain with the periodic boundary condition. The non-dimensional equations for fluid can be expressed as

$$\frac{\partial \hat{u}_i}{\partial t} + \hat{u}_j \hat{u}_{i,j} = -\hat{p}_{,i} + \frac{1}{\mathrm{Re}} \hat{u}_{i,jj} + Q\hat{u}_i - \hat{F}_{pi} \tag{4}$$

$$\hat{u}_{i,i} = 0 \tag{5}$$

$$\frac{\partial \hat{\theta}_f}{\partial t} + \hat{u}_j \frac{\partial \hat{\theta}_f}{\partial x_i} = -\frac{1}{\mathrm{Re}\,\mathrm{Pr}} \frac{\partial^2 \hat{\theta}_f}{\partial x_j^2} + \hat{u}_2 \bar{\nabla} T + \hat{q}_{2w} \tag{6}$$

The cap ' $\hat{}$ ' is used in Equations 4-6, indicating that the values used here are non-dimensionalized. This model, which is often called as homogeneous thermal convection model assumes that the temperature field can be decomposed into the fluctuating part $\hat{\theta}_f$ subjected to periodic boundary conditions and the constant mean part T_0.$\bar{\nabla} T$ in Equation 6 denotes the mean temperature gradient in the x_2 direction, which effectively acts as a source term for the fluid temperature field. The non-dimensional value of $\bar{\nabla} T$ is taken as 1.0 in the present simulations. Other parameters used in Equations 4, 5 and 6 are as follows: u is the velocity of the fluid, p is the pressure field, Re is the Reynolds number and Pr is the Prandtl number. Subscripts i and j represent tensor notations; and subscripts ',i' and ',j' represent differentiation with respect to x_i and x_j, respectively. Q is the linear forcing applied in the momentum equation to obtain a stationary isotropic turbulence. F_{pi}[23] in Equation 4 is the net force exerted by the particles on fluid and q_{2w}in Equation 6 is interfacial interaction between particles and liquid, which is modeled by addition of a temperature source term to the fluid temperature equation. It arises because of the convective heat transfer to and from the particle to fluid. In this model, q_{2w}acts as a coupling term to couple particle temperature source to the fluid temperature equation. This coupling term is calculated by applying the action-reaction principle to a generic volume of fluid (here considered as a grid cell) containing a particle. In this paper, the term $q_{2w}$is mentioned as a two-way temperature coupling term, and the effect of heat transport between particles and base fluid is called nanoparticle heat transfer. The equation for this coupling term is given as $q_{2w} = \sum_{n=1}^{N_p} \frac{\mathrm{Nu}}{2} \frac{\left(\theta_f(x^n) - T_p^n\right)}{\tau_T} \delta(x - x^n)$

While performing the simulations of thermal conductivity, fluid is initially considered to be at still condition and constant temperature of 300 K. Motion of fluid and change in fluid temperatures occur due to simultaneous interactions of particle dispersion and particle heat transport with the fluid medium. The value of Q is considered to be 0 for the simulations carried out to study the thermal conductivity of nanofluids. For the simulations considering the study of convective heat transfer, a stationary isotropic fluid state is obtained at Taylor's Reynolds number of 33.01. Taylor's Reynolds number is calculated using Taylor's microscale length as the characteristic length. Taylor's microscale length (λ) is the largest length scale at which fluid viscosity significantly affects the dynamics of turbulent eddies. Taylor's microscale length (λ) is given as $\lambda = (15\nu/\varepsilon)^{1/2}u'$, where ν is fluid viscosity, $\varepsilon$is fluid dissipation and u' is mean velocity fluctuations. Taylor's Reynolds number of 33.01 used in this simulation is equivalent to pipe flow Reynolds number of 5,500, and thus being turbulent, flow is chosen for this simulation. Simulating a higher Reynolds number at present is difficult due to an increase in thermal dissipation with an increase of Reynolds number, which will thus demand a very fine grid. The linear forcing coefficient used to maintain stationary turbulence is Q = 0.0667. The Prandtl number for all the simulations is taken as 5.1028, which is the Prandtl number of water at 300 K.

3. RESULTS

To validate the model, simulations were performed using the Cu(100 nm)/DIW **(distilled water)** and Al_2O_3(80 nm)/DIW nanofluids at different volume fractions. The turbulent thermal conductivity, which is the change in the conductivity of turbulent flow which is caused by the change of diffusivity of the flow, was determined by the equation $\langle u(x)\theta(x)\rangle = -k_T\bar{\nabla}T$ [24], where θ is the fluctuation of temperature. The effective thermal conductivity of the nanofluid was then calculated as k_{nf}/k_f = $(k_T + k_f)/k_f$, where k_f is the thermal conductivity of the fluid. The numerical data of present simulations is compared with the experimental data obtained by Xuan and Li [25] and Murshed et al. [26] (Figure 1). For the better understanding of the simulated results, values of the effective thermal conductivity of all the simulated nanofluids have been tabulated in Table 1. The calculated effective thermal conductivity values were observed to be in good agreement with the experimental data. The simulations underpredicted the effective thermal conductivity at 0.02 volume fraction for Cu(100 nm)/DIW nanofluid. A possible reason for this underprediction can be the discrepancy in prediction of the coagulation of particles in the present simulations, compared to the experiments. The values of effective thermal conductivity for the 0.03 and 0.05 volume fraction cases in the present simulations were closer to the experimental values. It can be observed that the values of Al_2O_3(80 nm)/DIW nanofluids show higher effective thermal conductivity at lower volume fractions in comparison with the effective thermal conductivity of Cu(100 nm)/DIW nanofluids. Cu(100 nm)/DIW nanofluids overtakes the effective thermal conductivity of Al_2O_3(80 nm)/DIW nanofluids at volume fraction above 0.02. Al_2O_3 being a non-metallic nanoparticle should have lower particle heat transport, which reduces the effectiveness of thermal conductivity enhancement at volume fraction greater than 0.02. However, at volume fractions lower than 0.02, higher effective thermal conductivity might be due to the smaller diameter of Al_2O_3 nanoparticles.

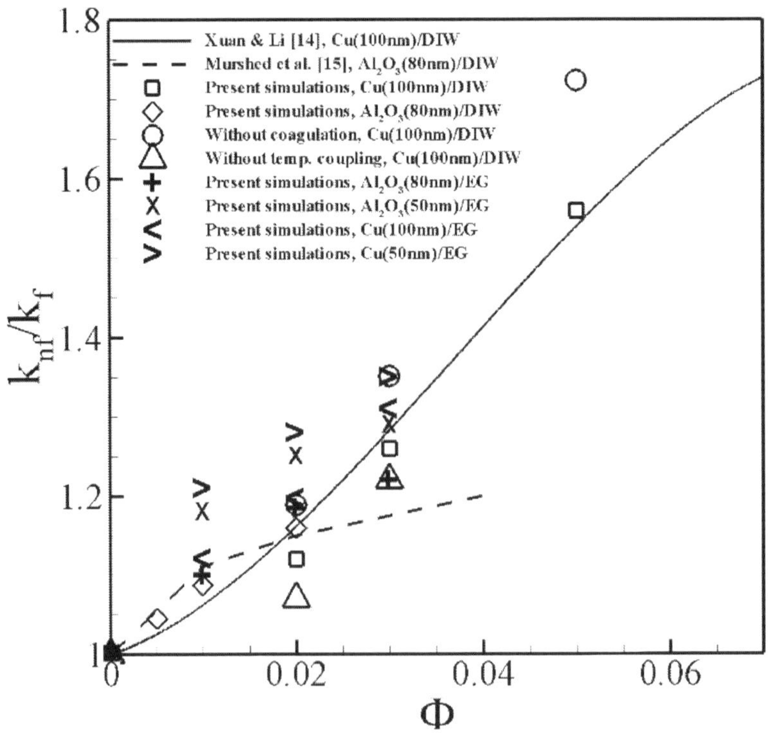

Figure 1. Effective thermal conductivity of nanofluids. Effective thermal conductivity of nanofluids at different volume fractions.

Table 1. Effective thermal conductivity of simulated nanofluids

Nanofluid Volume fraction	0.005	0.01	0.02	0.03	0.05
Cu(100 nm)/DIW			1.123	1.275	1.560
Cu(100 nm)/EG		1.135	1.191	1.313	
Cu(50 nm)/EG		1.220	1.273	1.362	
Al_2O_3(80 nm)/DIW	1.045	1.082	1.150		
Al_2O_3(80 nm)/EG		1.103	1.174	1.230	
Al_2O_3(50 nm)/EG		1.182	1.260	1.284	

Effective thermal conductivity of all simulated nanofluids is tabulated and shown here (computed values of effective thermal conductivity for simulations where the two-way temperature coupling and van der Waals force are neglected are not tabulated here).

In order to understand the effects of particle heat transport and coagulation of particles on thermal conductivity of nanofluids, simulations were performed for Cu(100 nm)/DIW nanofluids by neglecting two-way temperature coupling (q_{2w}) and van der Waals interaction force (F_{Vi}) one at a time. By neglecting two-way temperature coupling (q_{2w}), we forbid the contribution of particles to the heat transfer enhancement in nanofluids and only calculate the contribution of enhancement due to the dispersion of particle in the fluid medium. Similarly, by neglecting the van der Waals interaction force (F_{Vi}) we assume that the particles do not physically coagulate and observe the enhancement of heat transfer in nanofluids. Calculated effective thermal conductivity values are compared with the experimental data and simulation data where all the three parameters (i.e., particle dispersion, particle heat transport and coagulation of particles) are considered. When two-way temperature coupling is neglected, the results were found to be underpredicted by 4.45% for a 0.02-volume fraction of Cu(100 nm)/DIW nanofluid and by 3.62% for a 0.03-volume fraction of Cu(100 nm)/DIW nanofluid (Figure 1). The study suggests that both particle dispersions and particle heat transport have a contribution in the enhancement of effective thermal conductivity of nanofluids.

When the van der Waals force was neglected, the calculated thermal conductivity values are found to be overpredicted (Figure 1) as compared to experimental and simulation data where all the parameters are considered. Simulations, while neglecting the van der Waals force, were performed at 0.02, 0.03 and 0.05 volume fractions for Cu(100 nm)/DIW nanofluids. Overprediction of the calculated thermal conductivity is found to be increasing with an increase in the volume fraction. Difference between the calculated thermal conductivity values of with and without coagulation simulations is 6.13% for 0.02 volume fraction, 7.14% for 0.03 volume fraction and 10.47% for 0.05 volume fraction on Cu(100 nm)/DIW nanofluids. The study indicates that the coagulation of particles is one of the factors which are necessary to predict the thermal conductivity of nanofluids accurately.

Effect of different particle sizes and fluid medium on the effective thermal conductivity of nanofluids is also studied by performing simulations using Al_2O_3 nanoparticles of diameter 80 and 50 nm and Cu nanoparticles of diameter 100 and 50 nm by suspending them in the base fluid - EG. Simulations reveal that the size of nanoparticles has a great influence on the thermal conductivity of nanofluids. The smaller diameter of the particles will enhance the particle dispersion in the fluid medium which in turn can cause large disturbances in fluid and thus enhance the heat transfer rate of fluid. As can be seen from Figure 1 thermal conductivity of Al_2O_3 and Cu nanofluids increases dominantly when 50 nm particles are suspended in EG when in comparison with 80 or 100 nm particles. We have previously found that the decrease in size of nanoparticles leads to an increase in the particle dispersions and particle heat transport in the nanofluids which thus causes an increase in the effective thermal conductivity [18]. The figure also shows that with both DIW and EG base fluids, the thermal conductivity of nanofluids increases with increase in volume fraction. However, for a given volume fraction, it is observed that the thermal conductivity ratio enhancement is higher in EG. This behavior was consistently observed in both Cu and Al_2O_3 nanofluids. The reason for observed higher enhancement of thermal conductivity ratio in EG nanofluids could be due to the fact that the thermal conductivity of EG is low and thus the ratio of k_{nf}/k_f becomes larger.

The overall study of the thermal conductivity of nanofluids using the present model indicates a significant change in the effective thermal conductivity of nanofluids. Metallic nanoparticles were found to be more effective in enhancing the thermal conductivity of nanofluids. This could be due to stronger particle heat transport mechanism in metallic nanofluids. The study of different fluids indicates that nanoparticles, when suspended in EG, were more effective in enhancing the thermal conductivity of nanofluids. As the size of the nanoparticle decreases, the effective thermal conductivity of nanofluids was observed to be significantly enhanced. Simulations when performed by neglecting particle heat transport mechanism showed that the values of effective thermal conductivity are underpredicted, thus suggesting that both particle dispersion and particle heat transport have an effect on the enhancement of the effective thermal conductivity. Coagulation of particles is found to have a negative effect on the effective thermal conductivity enhancement. However, the simulations suggest that it is necessary to include van der Waals force in the numerical models to be able to accurately predict the thermal conductivity of nanofluids.

With the knowledge gained from the study of thermal conductivity of nanofluids, we included the terms particle dispersion, particle heat transport and coagulation of particles in our simulations of convective heat transfer in nanofluids. The study is more significant due to the fact that convective heat transfer of fluid has more practical applications. Also, though numerous simulations were performed to study the convective heat transfer enhancement in nanofluids, to our best knowledge the mechanism of heat transfer enhancement was not discussed by other researchers. We were interested in understanding the mechanism of heat transfer. An important question that lies ahead of us is if the particle dispersion of nanoparticles in fluid medium has a significant effect in the enhancement of the convective heat transfer in nanofluids.

In order to verify our model and also study the effect of different nanoparticle suspensions and size of nanoparticles on convective heat transfer of nanofluids, simulations were performed for Cu(100 nm)/DIW, Al_2O_3(100 nm)/DIW, CuO(100 nm)/DIW, TiO_2(100 nm)/DIW and SiO_2(100 nm)/DIW at 0.001, 0.005 and 0.01 volume fractions and for Cu(75 nm)/DIW, Cu(100 nm)/DIW and Cu(150 nm)/DIW at 0.005 volume fractions. The Nusselt number was calculated, using the formula $Nu = 1 + \frac{\langle u_2 \tilde{\nabla} T \theta_f \rangle}{\alpha}$, where α is the thermal diffusivity of fluid. The Nusselt number for Cu(100 nm)/DIW nanofluids at different volume fractions is compared with the experimental correlation (Figure 2) given in Xuan and Li [27] and is found to be in good agreement. The effect of volume fraction, particle material and particle size on the convective heat transfer can be observed in Figure 2. The Nusselt number increases with an increase in particle volume fraction and decreases with an increase in particle size. However, the enhancement of the Nusselt number is found to vary with the nanoparticle material suspended in the base fluid. For same volume fraction, it is found that the increase in Nusselt number is highest for Cu nanofluids and lowest for SiO_2 nanofluids. The difference in the enhancement of the Nusselt number for different particle materials is due to the difference in their particle heat transport in nanofluids. As explained below, the particle heat transport plays the most important role in enhancement of convective heat transfer in nanofluids. Simulations of Cu/DIW nanofluids at 0.005 volume fraction for different particle sizes were performed to understand the effect of different particle sizes on the convective heat transfer enhancement. Nusselt number of Cu/DIW nanofluids at 0.005 volume fraction for different particle sizes is shown in Figure 2 with open circle 'O' symbols. The effective Nusselt number of different simulated cases is tabulated and shown in Table 2. It can be observed that with an increase of particle size, the Nusselt number of nanofluids decreases.

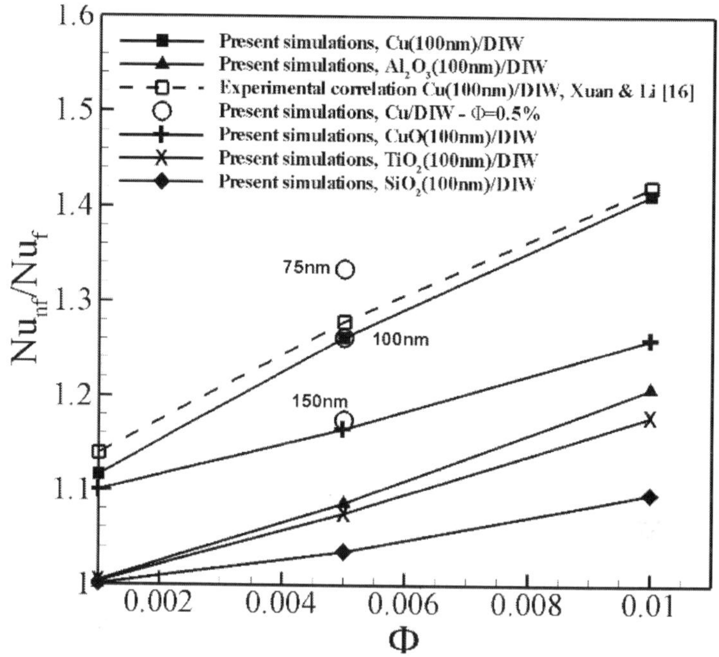

Figure 2. Effective Nusselt number of nanofluids. Effective Nusselt number for nanofluids at different volume fractions and particle diameters are shown.

Table 2. Effective Nusselt number of simulated nanofluids

Nanofluid Volume fraction	0.001	0.005	0.01
Cu(100 nm)/DIW	1.120	1.271	1.425
Al_2O_3(100 nm)/DIW	1.005	1.072	1.207
CuO(100 nm)/EG	1.100	1.161	1.259
TiO_2(100 nm)/DIW	1.003	1.067	1.187
SiO_2(100 nm)/EG	1.000	1.037	1.082
Cu(75 nm)/DIW		1.340	
Cu(150 nm)/DIW		1.164	

Effective Nusselt number of all simulated nanofluids is tabulated and shown here.

To understand the mechanism of convective heat transfer in turbulent nanofluids, distribution of the production terms ($P_{c\,2}$and $P_{c\,3}$) in transport equation of square temperature gradient (G_i^2) (Equation 7) and G_i^2 are plotted for Cu(100 nm)/DIW nanofluids at 0.001, 0.005 and 0.01 volume fractions (Figure 3). $P_{c\,1}$, which is production caused by the mean temperature gradient in fluid temperature equation (Equation 6) was found to be 70 times smaller compared to $P_{c\,2}$, which is production caused by the deformation of velocity field. Thus, it was assumed that the effect of $P_{c\,1}$on convective heat transfer is negligible and was not considered in further analysis. $P_{c\,3}$in Equation 7is production caused by the particle heat transport effect on fluid medium, which is represented as q_{2w}in Equation6. Distribution of G_i^2 shows an increase in the temperature gradients with an increase of particle volume fraction. However, the change in distribution of $P_{c\,2}$with change in particle volume fraction is found to be negligible. It suggests that the particle dispersions, which deform the fluid velocity, do not significantly affect the convective heat transfer rate in nanofluids. On the other hand, distribution of $P_{c\,3}$shows a significant difference at different particle volume fractions. Moreover, the high temperature gradients are found to be distributed in the regions of high magnitudes of $P_{c\,3}$. It suggests a significant influence of particle heat transport on convective heat transfer of nanofluids.

$$\frac{\partial}{\partial t}\left(\frac{1}{2}G_i^2\right) = \underbrace{-\frac{1}{2}S_{\theta}G_j u_{i,j}}_{P_{c1}} \underbrace{-G_i G_j S_{ij}}_{P_{c2}} \underbrace{+\alpha\left(\frac{\partial G_i}{\partial x_j}\right)^2}_{\text{Dissipation}} \underbrace{-\alpha\frac{\partial^2}{\partial x_i^2}\left(\frac{1}{2}G_i^2\right)}_{\text{Diffusion}} + \underbrace{(\text{Extra term due to particles})}_{P_{c3}} \quad (7)$$

Simulations performed to study the convective heat transfer in nanofluids reveal that the convective heat transfer in nanofluids has significant influence from the kind of nanoparticles suspended in fluid medium. It was observed that the nanoparticles with higher heat transport rate show more enhancements in Nusselt number of nanofluids. The study of square temperature gradient and its production terms indicates that Equation 7, reveals that the particle dispersions in turbulent fluid, unlike in still fluid, do not significantly affect the heat transfer rate. It can be due to the presence of a large drag force on particles when the fluid is under turbulent conditions. The presence of a large drag force on particles in moving fluid nullifies the effect of other forces such as the Brownian force and thermophoresis force. However, all the simulations performed for the study of convective heat transport phenomenon in this paper, due to computational limitations, use nanoparticles with size 100 nm. We therefore have to study the effect of particle dispersions on convective heat transfer of nanofluids while using smaller sized particles, before a foregone conclusion can be made on the effect of particle dispersions.

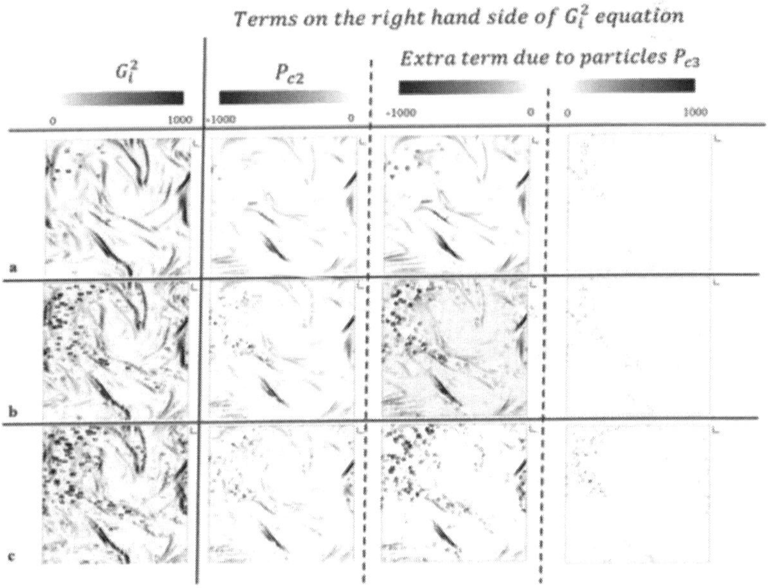

Figure 3. Distribution of terms in square temperature gradient. Distribution of G_i^2 , $P_{c\ 2}$and negative and positive terms of $P_{c\ 3}$are shown for Cu(100 nm)/DIW nanofluids at (a) Φ = 0.001, (b) Φ = 0.005 and (c) Φ = 0.01. Reprint from S. Kondaraju, E. K. Jin and J. S. Lee, Investigation of heat transfer in turbulent nanofluids using direct numerical simulations, 81, 016304, 2010. "Copyright 2010 by the American Physical Society."

4. CONCLUSIONS

In this study, we have made an attempt to present a numerical model which can simulate and predict the thermal conductivity and also convective heat transfer in nanofluids. The model showed a good agreement with the experimental data. A wide range of particle sizes and nanoparticle materials used in the study also agree qualitatively with the results of previous researchers. A significant advantage of the present study is that it can help in understanding the mechanism of enhancement of thermal conductivity and Nusselt number in nanofluids.

ACKNOWLEDGEMENTS

This work was partially supported by grants from Basic Science Research Program through the National Research Foundation of Korea (NRF) funded by the Ministry of Education, Science and Technology (grant number, 2010-0007113) and Brain Korea (BK) 21 HRD Program for Nano Micro Mechanical Engineering.

APPENDIX

If the diameter of the two particles is considered as d_1 and d_2, an increase in the volume of particles (due to the method of coagulation in the present model) in the computational domain due to the agglomeration of two particles is given as follows.

$$\text{Increase in volume of particles} = \frac{\pi(d_1+d_2)^3}{6} - \left(\frac{\pi\left(d_1^3+d_2^3\right)}{6}\right) = 3\left(d_1^2 d_2 + d_1 d_2^2\right)$$

The maximum increase in the volume of particles in the computational domain will be observed when all the particles coagulate into one single particle. The maximum number of particles

(n) used in this study is 500,000 and the largest diameter of particles used is 100 nm. Thus, the maximum increase of volume of particles due to the present coagulation model is

$$\text{Maximum increase in the volume of particles} = 3\left((d_1 \times n)^2 d_1 + (d_1 \times n) d_1^2\right)$$

When $n = 500{,}000$ and $d_1 = 100$ nm,

The maximum increase in the volume of particles approximately equal to 15×10^{-11}

Thus, it can be observed that the increase in the volume concentration of particles due to the present coagulation model will have a negligible effect on the simulated results.

AUTHORS' CONTRIBUTIONS

SK has carried out the simulations and participated in the analysis and interpretation of data. He also participated in drafting the manuscript. JSL conceived in the study and participated in data analysis and drafted the manuscript.

REFERENCES

1. Maxwell JC: *Treatise on Electricity and Magnetism.* Oxford: Clarendon Press; 1873.

2. Ahuja AS: Augmentation of heat transfer in laminar flow of polystyrene suspensions. *J Appl Phys* 1975, 46: 3408–3425. 10.1063/1.322107

3. Liu KV, Choi SUS, Kasza KE: Measurements of pressure drop and heat transfer in turbulent pipe flows of particulate slurries.*ANL-88–15, Argonne National Laboratory Report* 1988.

4. Eastman JA, Choi SUS, Li S, Soyez G, Thompson LJ, Melfi RJD: Novel thermal properties of nanostructured materials. *Mater Sci Forum* 1999, 312–314: 629–634. 10.4028/www.scientific.net/MSF.312-314.629

5. Choi SUS: *Developments Applications of Non-Newtonian Flows.* New York: ASME; 1995.

6. Eastman JA, Choi SUS, Li S, Thompson LJ: Enhanced thermal conductivity through the development of nanofluids. In*Proceedings of the Symposium on Nanophase and Nanocomposite Materials II. Volume 457.* Materials Research Society, Boston, USA; 1997:3–11.

7. Wang X, Xu X, Choi SUS: Thermal conductivity of nanoparticle-fluid mixture. *J Thermophys Heat Transfer* 1999, 13: 474–480.

8. Choi SUS, Zhang ZG, Yu W, Lockwood FE, Grulke EA: Anomalous thermal conductivity enhancement in nanotube suspensions. *Appl Phys Lett* 2001, 79: 2252–2254.

9. Murshed SMS, Leong KC, Yang C: Investigations of thermal conductivity and viscosity of nanofluids. *Int J Thermal Sci* 2008, 47: 560–568.

10. Li CH, Peterson GP: Experimental investigation of temperature and volume fraction variations on the effective thermal conductivity of nanoparticle suspensions (nanofluids). *J Appl Phys* 2006, 99: 084314.

11. Zhu H, Zhang C, Liu S, Tang Y, Yin Y: Effects of nanoparticles clustering and alignment on thermal conductivities of Fe3O4 aqueous nanofluids. *Appl Phys Lett* 2006, 89: 023123.

12. Bhattacharya P, Saha SK, Yadav A, Phelan PE, Prasher RS: Brownian dynamics simulation to determine the effective thermal conductivity of nanofluids. *J Appl Phys* 2004, 95: 6492.

13. Jain S, Patel HE, Das SK: Brownian dynamics simulation for the prediction of effective thermal conductivity of nanofluids. *J Nanopart Res* 2009, 11: 767–773.

14. Xuan Y, Yao Z: Lattice Boltzmann model for nanofluids. *Heat Mass Trans* 2005, 41: 199–205.

15. Evans W, Fish J, Keblinski P: Role of Brownain motion hydrodynamics on nanofluid thermal conductivity. *Appl Phys Lett* 2006, 88: 093116.

16. Sarkar RS, Selvam P: Molecular dynamics simulation of effective thermal conductivity and study of enhanced thermal transport mechanism in nanofluids. *J Appl Phys* 2007, 102: 074302.

17. Maiga SEB, Nguyen CT, Galanis N, Roy G: Heat transfer behaviors of nanofluids in a uniformly heated tube. *Superlatt Microstructure* 2004, 35: 543–557.

18. Kondaraju S, Lee JS: Effect of the multi-sized nanoparticle distribution on the thermal conductivity of nanofluids. *Microfluid Nanofluid* 2010, 10: 133–144.

19. Maxey MR, Riley JJ: Equation of motion for a small rigid sphere in a uniform flow. *Phys Fluids* 1983, 26: 883–890.

20. Li A, Ahmadi G: Dispersion and deposition of spherical particles from point sources in a turbulent channel flow. *Aerosol Sci Tech* 1992, 16: 209–226.

21. Talbot L, Cheng RK, Schefer RW, Willis DR: Thermophoresis of particles in a heated boundary layer. *J Fluid Mech* 1980, 101: 737–758.

22. Apostulou K, Hyrmak AN: Discrete element simulation of particle-laden flow. *Comput Chem Eng* 2008, 32: 841–856.

23. Sundaram S, Collins LR: A numerical study of the modulation of isotropic turbulence by suspended particles. *J Fluid Mech*1999, 379: 105–143.

24. Elperin T, Kleeorin N, Rogachevkii I: Turbulent thermal diffusion of small inertial particles. *Phys Rev Lett* 1996, 76: 224–227.

25. Xuan Y, Li Q: Heat transfer enhancement of nanofluids. *Int J Heat Fluid Flow* 2000, 21: 58–64.

26. Murshed SMS, Leong KC, Yang C: Thermophysical and electrokinetic properties of nanofluids-A critical review. *Int J Nanoscience* 2006, 5: 23.

27. Xuan Y, Li Q: Investigation on convective heat transfer and flow features of nanofluids. *J Heat Transfer* 2003, 125: 151–156.

INDEX